Contents

Introduction

■ ■ ■

Content Guidance

■ ■ ■

Questions and Answers

Introduction

About this guide

This unit guide is one of a series covering the Edexcel specification for AS and A2 chemistry. It offers advice for the effective study of **Unit 5: Transition Metals and Organic Nitrogen Chemistry**. Its aim is to help you *understand* the chemistry. It is not intended as a shopping list, enabling you to cram for an examination. The guide has three sections:

- **Introduction** — this provides guidance on study and revision, together with advice on approaches and techniques to ensure you answer the examination questions in the best way that you can.

- **Content Guidance** — this section is not intended to be a textbook. It offers guidelines on the main features of the content of Unit 5, together with particular advice on making study more productive.

- **Questions and Answers** — this section shows you the sort of questions you can expect in the unit test. Answers are provided; in some cases, distinction is made between responses that might have been given by a grade-A candidate and typical errors that are often made. Careful consideration of these will improve your answers and, more importantly, will increase your understanding of the chemistry involved.

The effective understanding of chemistry requires time. No one suggests it is an easy subject, but even those who find it difficult can overcome their problems by the proper investment of time.

To understand the chemistry, you have to make links between the various topics. The subject is coherent; it is not a collection of discrete modules. These links come only with experience, which means time spent thinking about chemistry, working with it and solving chemical problems. Time produces fluency with the ideas. If you have that, together with good technique, the examination will look after itself.

The specification

The specification states the chemistry that can be examined in the unit tests and describes the format of the tests. This is not necessarily the same as what teachers might choose to teach or what you might choose to learn.

The purpose of this book is to help you with Unit Test 5, but don't forget that what you are doing is learning *chemistry*. The specification can be obtained from Edexcel, either as a printed document or from the web at www.edexcel.com.

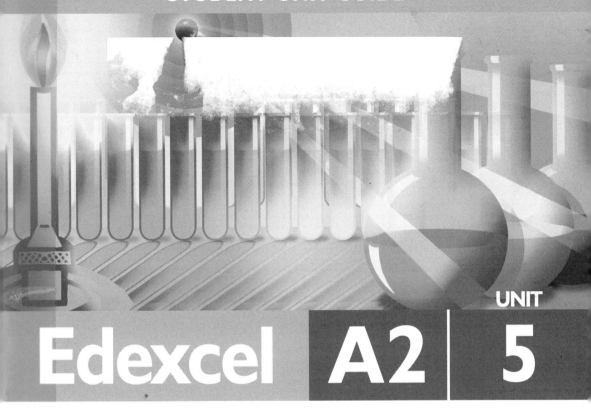

UNIT

Edexcel A2 | 5

Chemistry

Transition Metals and
Organic Nitrogen Chemistry

George Facer

Philip Allan Updates, an imprint of Hodder Education, an Hachette UK company, Market Place, Deddington, Oxfordshire OX15 0SE

Orders
Bookpoint Ltd, 130 Milton Park, Abingdon, Oxfordshire OX14 4SB
tel: 01235 827827
fax: 01235 400401
e-mail: education@bookpoint.co.uk

Lines are open 9.00 a.m.–5.00 p.m., Monday to Saturday, with a 24-hour message answering service. You can also order through the Philip Allan Updates website: www.philipallan.co.uk

© Philip Allan Updates 2009

ISBN 978-0-340-94949-8

First printed 2009
Impression number 5 4 3
Year 2014 2013 2012 2011

This guide has been written specifically to support students preparing for the Edexcel A2 Chemistry Unit 5 examination. The content has been neither approved nor endorsed by Edexcel and remains the sole responsibility of the author.

Typeset by Tech-Set Ltd, Gateshead, Tyne and Wear
Printed by MPG Books, Bodmin

Hachette UK's policy is to use papers that are natural, renewable and recyclable products and made from wood grown in sustainable forests. The logging and manufacturing processes are expected to conform to the environmental regulations of the country of origin.

The unit test

The Unit 5 test lasts 1 hour 40 minutes and has three sections:
- **Section A** — about 20 multiple-choice questions, each with a choice of four answers.
- **Section B** — a mixture of short-answer and extended-answer questions. This section resembles the papers in the previous modular tests, where the more extended writing is worth about 4 or 5 marks.
- **Section C** — the questions are based on a contemporary topic contained in a passage that you must read. Some of the questions require understanding of parts of the passage and some deal with topics similar to that in the passage. Section C is not a comprehension exercise, but tests the chemistry behind the topics in the passage.

Command terms

Examiners use certain words that require you to respond in a specific way. You must distinguish between these terms and understand exactly what each requires you to do.
- **Define** — give a simple definition without any explanation.
- **Identify** — give the name or formula of the substance.
- **State** — no explanation is required (and you shouldn't give one).
- **Deduce** — use the information supplied in the question to work out your answer.
- **Suggest** — use your knowledge and understanding of similar substances, or those with the same functional groups, to work out the answer.
- **Compare** — make a statement about *both* substances being compared.
- **Explain** — use chemical theories or principles to say why a particular property is as it is.
- **Predict** — say what you think will happen on the basis of the principles that you have learned.
- **Justify your answer** — give a brief explanation of why your answer is correct.

Calculations

You must show your working in order to score full marks. Be careful about significant figures. If a question does not specify the number of significant figures required, give your answer to *three significant figures* or to two decimal places for pH calculations.

Organic formulae

- **Structural formula** — you must give a structure that is unambiguous. For instance, $CH_3CH_2CH_2OH$ or $C_2H_5CH_2OH$ are acceptable, but C_3H_7OH could be either propan-1-ol or propan-2-ol and so is not acceptable. If a compound has a double bond, then it should be shown in the structural formula.
- **Displayed or full structural formula** — you must show all the *atoms* and *all* the *bonds*. 'Sticks' instead of hydrogen atoms will lose marks.

- **Shape** — if the molecule or ion is pyramidal, tetrahedral or octahedral you must make sure that your diagram looks three-dimensional by using wedges and dashes. Draw optical isomers as mirror images of each other. Geometric isomers must be drawn with bond angles of 120°. Make sure that the *bonds go to the correct atom* — for example, the oxygen in an —OH group or the carbon in —CH_3 and —COOH groups.

Points to watch

- **Stable** — if you use this word, you must qualify it — for example, 'stable *to heat*', 'the reaction is *thermodynamically* stable', 'the reaction is *kinetically* stable' or 'a secondary carbocation intermediate is *more* stable *than* a primary carbocation'.
- **Reagents** — if you are asked to identify a reagent, you must give its *full* name or formula. Phrases such as 'acidified dichromate(VI)' will not score full marks. You must give the reagent's full name — for example, 'potassium dichromate(VI)'.
- **Conditions** — don't use abbreviations such as 'hur' for heat under reflux.
- **Atoms, molecules and ions** — don't use these words randomly. Ionic compounds contain ions, not molecules.
- **Rules** — don't use rules such as Markovnikov or Le Chatelier to *explain*. However, they can be used to predict.
- **Melting and boiling** — when a molecular covalent substance (such as water) is melted or boiled, *covalent* bonds are *not* broken. So melting and boiling points are connected with the type and strength of *intermolecular* forces. When an ionic substance is melted, the ionic bonds are *not* broken — the substance is still ionic. The ions gain enough energy to separate.

Learning to learn

Learning is not instinctive — you have to develop suitable techniques to make good use of your time. In particular, chemistry has peculiar difficulties that need to be understood if your studies are to be effective from the start.

Planning

Busy people do not achieve what they do haphazardly. They plan — so that if they are working they mean to be working, and if they are watching television they have planned to do so. Planning is essential. You must know what you have to do each day and each week and set aside time to do it.

Be realistic in your planning. You cannot work all the time, so you must build in time for recreation and family responsibilities.

introduction

Targets

When devising your plan, have a target for each study period. This might be a particular section of the specification, or it might be rearranging of information from text into pictures, or the construction of a flowchart relating all the organic reactions you need to know. Whatever it is, be determined to master your target material before you leave it.

Reading chemistry textbooks

Chemistry textbooks are a valuable resource, not only for finding out the information for your homework but also to help you understand concepts of which you are unsure. They need to be read carefully, with a pen and paper to hand for jotting down things as you go — for example, making notes, writing equations, doing calculations and drawing diagrams. Reading and revising are *active* processes that require concentration. Looking vaguely at the pages is a waste of time. In order to become fluent and confident in chemistry, you need to master detail.

Chemical equations

Equations are quantitative, concise and internationally understood. When you write an equation, check that:

- you have thought of the *type* of reaction occurring — for example, is it neutralisation, addition or disproportionation?
- you have written the correct formulae for all the substances
- your equation balances both for the numbers of atoms of each element and for charge
- you have not written something silly, such as having a strong acid as a product when one of the reactants is an alkali
- you have included *state symbols* in all thermochemical equations and if they have been asked for

Graphs

Graphs give a lot of information, and they must be understood in detail rather than as a general impression. Take time over them. Note what the axes are, the units, the shape of the graph and what the shape means in chemical terms. Think about what could be calculated from the graph. Note whether the graph flattens off and what that means. This is especially important in kinetics.

When drawing a graph, do not join up the points — draw a smooth line (straight or curved) as near as possible to all the points. However, if you are plotting a list, such as the first ionisation energies of the elements, you do join up the points.

Tables

These are a means of displaying a lot of information. You need to be aware of the table headings and the units of numerical entries. Take time over them. What trends can be seen? How do these relate to chemical properties? Make sure that you use all the data when answering an examination question.

Diagrams

Diagrams of apparatus should be drawn in section. When you see them, copy them and ask yourself why the apparatus has the features it has. What is the difference between a distillation and a reflux apparatus, for example? When you do practical work, examine each piece of the apparatus closely so that you know both its form and function.

Calculations

Calculations are not normally structured in A2 as they were in AS. Therefore, you will need to *plan* the procedure for turning the data into an answer.

- Set your calculations out fully, making it clear what you are calculating at each step. Don't round figures up or down during a calculation. Either keep all the numbers on your calculator or write any intermediate answers to four significant figures.
- If you have time, check the accuracy of each step by recalculating it. It is so easy to enter a wrong number into your calculator or to calculate a molar mass incorrectly.
- Think about the size of your final answer. Is it far too big or vanishingly small?
- Finally, check that you have the correct *units* in your answer and that you have given it to an appropriate number of *significant figures* — if in doubt, give it to three or, for pH calculations, to two decimal places.

Notes

Most students keep notes of some sort. Notes can take many forms: they might be permanent or temporary; they might be lists, diagrams or flowcharts. You have to develop your own styles. For example, notes that are largely words can often be recast into charts or pictures, which is useful for imprinting the material. The more you rework the material, the clearer it will become.

Whatever form your notes take, they must be organised. Notes that are not indexed or filed properly are useless, as are notes written at enormous length and those written so cryptically that they are unintelligible a month later.

Writing

There is some requirement for extended writing in Unit Test 5. You need to be able to write concisely and accurately. This requires you to marshal your thoughts properly and needs to be practised during your ordinary learning.

For experimental plans, it is a good idea to write your answer as a series of bullet points. There are no marks specifically for 'communication skills', but if you are not able to communicate your ideas clearly and accurately, then you will not score full marks. The space available for an answer is a poor guide to the amount that you have to write — handwriting sizes differ hugely, as does the ability to write crisply. Filling the space does not necessarily mean you have answered the question. The mark allocation suggests the number of points to be made, not the amount of writing needed.

Approaching the unit test

The unit test is designed to allow you to show the examiner what you know. Answering questions successfully is not only a matter of knowing the chemistry but is also a matter of technique. All questions in Unit Test 5 are answered on the question paper.

Revision

- Start your revision in plenty of time. Make a list of what you need to do, emphasising the topics that you find most difficult and draw up a detailed revision plan. Work back from the examination date, ideally leaving an entire week free from fresh revision before that date. Be realistic in your revision plan and then add 25% to the timings because everything takes longer than you think.
- When revising, make a note of difficulties and ask your teacher about them. If you do not make these notes, then you will forget to ask.
- Make use of past papers. Similar questions are regularly asked, so if you work through as many past papers and answers as possible, then you will be in a strong position to obtain a top grade.
- When you use the Question and Answer section of this guide, make a determined effort to write your answers *before* looking at the sample answers and examiner comments.

The exam

- Read the question. Questions usually change from one examination to the next. A question that looks the same, at a cursory glance, to one that you have seen before, usually has significant differences when read carefully. Needless to say, candidates do not receive credit for writing answers to their own questions.
- Be aware of the number of marks available for a question. This is an excellent pointer to the number of things you need to say.
- Do not repeat the question in your answer. The danger is that you fill up the space available and think that you have answered the question, when in reality some or maybe all of the real points have been ignored.
- Look for words in **bold** in a question and make sure that you have answered the question fully in terms of those words or phrases. For example, if the question asks you to define the **standard electrode potential**, make sure that you explain the meaning of electrode potential as well as what the standard conditions are.

- Be careful in negative multiple-choice questions. The word 'NOT' will be in upper-case letters or in **bold**.
- Questions in Unit Test 5 will often involve substances or situations that are new to you. This is deliberate and is what makes these questions synoptic. Don't be put off by large organic molecules. They are nothing more than a collection of functional groups that you can assume react independently of each other.

Unit Test 5 has three assessment objectives:
- AO1 is 'knowledge with understanding of science and how science works' and makes up 30% of the test. You should be able to:
  - recognise, recall and show understanding of specific chemical facts, principles, concepts, practical techniques and terminology
  - select, organise and communicate information clearly and logically, using specialist vocabulary where appropriate
- AO2 is 'application of knowledge and understanding of science and how science works' and makes up 54.5% of Unit Test 5. You should be able to:
  - analyse and evaluate scientific knowledge and processes
  - apply scientific knowledge and processes to unfamiliar situations including those related to issues
  - assess the validity, reliability and credibility of scientific information
- AO3 is 'How Science Works' and makes up 15.5% of the test. You should be able to:
  - describe ethical, safe and skilful practical techniques and processes, selecting appropriate qualitative and quantitative methods
  - analyse, interpret and explain and evaluate the methodology, results and impact of experimental and investigative activities

Content
Guidance

This section is a guide to the content of **Unit 5: Transition Metals and Organic Nitrogen Chemistry**. It does not constitute a textbook for the material in Unit 5.

The topics covered in this section are:
- Redox
- Transition-metal chemistry
- Organic chemistry: arenes
- Organic chemistry: nitrogen compounds
- Organic chemistry: analysis and synthesis
- Organic reactions: summary

Redox

Required AS chemistry

Questions set on this topic may require the application of knowledge of redox from Unit 2.

Definitions

- **Oxidation** occurs when an atom, molecule or ion *loses* one or more electrons. The oxidation number of the element involved increases.
- **Reduction** is when an atom, molecule or ion *gains* one or more electrons. The oxidation number of the element involved decreases.
- An **oxidising agent** is a substance that oxidises another substance and is itself reduced. The oxidation number of an element in the oxidising agent decreases.
- A **reducing agent** is a substance that reduces another substance and is itself oxidised. Its oxidation number increases.
- The **oxidation number** of an atom in a compound (or ion) is the charge that it would have if the compound (or ion) were ionic.

> **Tip** Remember **OILRIG** — **O**xidation **I**s **L**oss, **R**eduction **I**s **G**ain.

Oxidation number

The rules for working out oxidation numbers should be applied in the following order:
- **Rule 1** — the oxidation number of an element in its standard state is zero.
- **Rule 2** — a simple monatomic ion has an oxidation number equal to its charge.
- **Rule 3** — the sum of the oxidation numbers in a neutral formula is zero.
- **Rule 4** — the sum of the oxidation numbers in an ion adds up to the charge on that ion.
- **Rule 5** — the oxidation number of hydrogen is +1, except in metal hydrides where it is −1.
- **Rule 6** — the oxidation number of oxygen is −2, except in peroxides (e.g. hydrogen peroxide, H_2O_2, where it is −1) or when combined with fluorine, where it is +2.

Worked example

What is the oxidation number of iron in Fe_3O_4?

Answer

The oxidation state of each oxygen is −2 (rule 6); therefore, the four oxygen atoms are $4 \times -2 = -8$. The three iron atoms are +8 in total (rule 3). Thus, the average oxidation state of each iron atom is $8 \div 3 = 2.667$ $(2\frac{2}{3})$.

The answer is not a whole number because two of the iron atoms are in the +3 state and one is in the +2 state.

Disproportionation

This is a reaction in which an *element* in a *single* species is simultaneously oxidised and reduced. In the following reaction, the chlorine in the species $NaOCl$ is oxidised from $+1$ to $+5$ and simultaneously reduced from $+1$ to -1:

$$3NaOCl \rightarrow NaClO_3 + 2NaCl$$
$$\quad\; +1 \qquad\;\; +5 \qquad\qquad -1$$

The following reaction is *not* a disproportionation reaction, even though the chlorine in Cl^- has been oxidised and the chlorine in OCl^- has been reduced, because the chlorine is in two *different* species:

$$Cl^- + OCl^- + 2H^+ \rightarrow Cl_2 + H_2O$$
$$-1 \quad\;\; +1 \qquad\qquad\quad 0$$

A2 chemistry

Change in oxidation number and reaction stoichiometry

The increase in oxidation number of the element in the reducing agent must equal the decrease in oxidation number of the element in the oxidising agent. Note that a reducing agent gets *oxidised* and an *oxidising* agent gets reduced

Worked example

Bromate ions (BrO_3^-) react with bromide ions (Br^-), in acidic solution to form bromine and water. Evaluate the oxidation numbers of the bromine species and hence work out the reaction stoichiometry. Write the overall equation.

Answer

The oxidation numbers in BrO_3^- add up to -1 (rule 4). Each oxygen is -2 (rule 6), so the three oxygen atoms are $3 \times -2 = -6$. The bromine in BrO_3^- is $+5$, because $+5 + (-6) = -1$.

The oxidation number of bromine in Br^- is -1 (rule 2).

The oxidation number of bromine in Br_2 is zero (rule 1).

As the oxidation number of bromine in BrO_3^- decreases by five and the bromine in Br^- increases by one, there must be $5Br^-$ to each BrO_3^-.

Therefore the equation is:

$$5Br^- + BrO_3^- + 6H^+ \rightarrow 3Br_2 + 3H_2O$$

Tip If the question mentions 'in acidic solution', there will be H^+ ions on the left in the overall equation.

Standard electrode potential

This is defined as the electrical potential of an *electrode* relative to a standard hydrogen electrode, where all solutions are at $1\,mol\,dm^{-3}$ concentration and any gases are at 1 atm pressure and a stated (usually 25°C) temperature.

The values are always stated as *reduction* potentials, with electrons on the left:

$$Zn^{2+}(aq) + 2e^- \rightleftharpoons Zn(s) \qquad E^{\ominus} = -0.76\,V$$

so the standard electrode potential is also called the standard *reduction* potential. The data can be given as a reduction equation, as above, or as $E^{\ominus}\,(Zn^{2+}/Zn) = -0.76\,V$.

Standard hydrogen electrode

This is the reference electrode in all electrochemical measurements. Its value at 25°C is defined as zero volts. It consists of hydrogen gas at 1 atm pressure being passed over a platinum electrode dipping into a solution of $1.0\,mol\,dm^{-3}$ H^+ ions (pH = 0) at a stated temperature:

$$H^+(aq) + e^- \rightleftharpoons \tfrac{1}{2}H_2(g) \qquad E^{\ominus} = 0.00\,V$$

A reference electrode is needed, as a potential cannot be measured on its own — only a potential difference.

Measurement of standard electrode potentials

The voltage of the electrode or half-cell is measured relative to a standard, such as the standard hydrogen electrode. For zinc, the half-equation is $Zn^{2+}(aq) + 2e^- \rightleftharpoons Zn(s)$. Thus, it is the potential of a zinc rod dipping into a $1.0\,mol\,dm^{-3}$ solution of Zn^{2+} ions at 298 K, relative to a standard hydrogen electrode.

For chlorine, the half-equation is $Cl_2(g) + 2e^- \rightleftharpoons 2Cl^-(aq)$. Thus, it is the potential of chlorine gas, at 1 atm pressure, bubbled over a platinum electrode dipping into a $1.0\,mol\,dm^{-3}$ solution of Cl^- ions at 298 K, relative to a standard hydrogen electrode.

For the Fe^{3+}/Fe^{2+} system, the half-equation is $Fe^{3+}(aq) + e^- \rightleftharpoons Fe^{2+}(aq)$. Thus, it is the potential when a platinum electrode is placed into a solution of concentration $1.0\,mol\,dm^{-3}$ in *both* Fe^{3+} and Fe^{2+} ions at 298 K, relative to a standard hydrogen electrode.

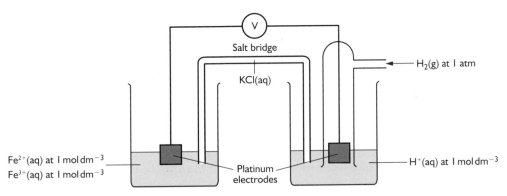

Note that all soluble substances, whether they are on the left or the right of the half-equation, must be at a concentration of $1.0 \, \text{mol dm}^{-3}$, if the standard potential is being measured.

Feasibility and extent of reaction

A reaction is said to be feasible (thermodynamically spontaneous) if the value of the cell potential ($E_{cell}^{\ominus}$) is *positive*. The extent of the reaction depends on the number of electrons involved and the numerical value of $E_{cell}^{\ominus}$. As a rough guide, reactions with an $E_{cell}^{\ominus} > 0.3 \, \text{V}$ will be virtually complete.

The total entropy change, ΔS_{total}, is directly proportional to the value of $E_{cell}^{\ominus}$:

$$\Delta S_{total} = nE_{cell}^{\ominus} \frac{F}{T}$$

where n is the number of electrons exchanged and F is the Faraday constant $= 96\,500 \, \text{C}$ and T the temperature in kelvin.

For a reaction with $E_{cell}^{\ominus} = +0.3 \, \text{V}$ and two electrons involved:

$$\Delta S_{total} = 2 \times 0.3 \times \frac{96\,500}{298} = 194 \, \text{J K}^{-1} \, \text{mol}^{-1}$$

ΔS_{total} is connected with the equilibrium constant by the expression:

$$\Delta S_{total} = R \ln K$$

where R is the gas constant $= 8.31 \, \text{J K}^{-1} \, \text{mol}^{-1}$, so:

$$K = e^{\frac{194}{8.31}} = 1.4 \times 10^{10} \text{ at } 298 \, \text{K} \, (25°\text{C})$$

Calculation of $E_{cell}^{\ominus}$ for a reaction and feasibility

If a question asks 'Will substance A oxidise substance B?', you need to look for the half-equations containing A and B. You will find A on the left and B on the right of their half-equations.

There are two methods for calculating $E_{cell}^{\ominus}$ (which is also called $E_{reaction}^{\ominus}$).

Method 1

This is based on the formula:

$$E_{cell}^{\ominus} = E^{\ominus} \text{(oxidising agent)} - E^{\ominus} \text{(reducing agent)}$$

where $E^{\ominus}$ refers to the standard reduction potential, i.e. the half-equation with electrons on the *left*.

- **Step 1:** Identify the oxidising agent. Remember that the oxidising agent is reduced and so is on the left of a standard electrode (reduction) potential equation.

- **Step 2**: Identify the reducing agent. The reducing agent is oxidised and so is on the right of a standard electrode (reduction) potential.
- **Step 3**: Calculate $E^{\ominus}_{cell} = E^{\ominus}$ of oxidising agent $- E^{\ominus}$ of reducing agent, where the $E^{\ominus}$ values are the standard *reduction* potentials of the oxidising agent (on the left in its half-equation) and of the reducing agent (on the right in its half-equation).

Thus, a reaction will be feasible if the $E^{\ominus}$ (oxidising agent) is more positive (or less negative) than the $E^{\ominus}$ of the half-equation with the reducing agent on the right. The bigger the positive value of $E^{\ominus}$, the stronger the substance is as an oxidising agent.

Worked example 1

Use the following data to decide whether dichromate(VI) ions oxidise chloride ions in acid solution:

$$Cr_2O_7^{2-} + 6e^- + 14H^+ \rightleftharpoons 2Cr^{3+} + 7H_2O \qquad E^{\ominus} = +1.33\,V$$
$$Cl_2 + 2e^- \rightleftharpoons 2Cl^- \qquad E^{\ominus} = +1.36\,V$$

Answer

The reactants are $Cr_2O_7^{2-}$ and Cl^-.

Step 1: The oxidising agent is $Cr_2O_7^{2-}$. It is reduced to Cr^{3+}.

Step 2: The reducing agent is Cl^-. It is oxidised to Cl_2.

Step 3: $E^{\ominus}_{cell} = E^{\ominus}(Cr_2O_7^{2-}/Cr^{3+}) - E^{\ominus}(Cl_2/Cl^-)$
$$= +1.33 - (+1.36) = -0.03\,V$$

$E^{\ominus}_{cell}$ is *negative* and so the reaction does *not* take place.

Worked example 2

Will manganate(VII) ions in acid solution oxidise chloride ions to chlorine?

$$Cl_2 + 2e^- \rightleftharpoons 2Cl^- \qquad E^{\ominus} = +1.36\,V$$
$$MnO_4^- + 8H^+ \rightleftharpoons Mn^{2+} + 4H_2O \qquad E^{\ominus} = +1.52\,V$$

Answer

The manganate(VII) ions have a larger positive $E^{\ominus}_{cell}$ than chlorine, so they are the oxidising agent.

The reducing agent is the species on the *right* of the half-equation with the lower $E^{\ominus}$ value. Thus, chloride ions are the reducing agent.

$E^{\ominus}_{cell} = E^{\ominus}(MnO_4^-/Mn^{2+}) - E^{\ominus}(Cl_2/Cl^-) = +1.52 - (+1.36) = +0.16\,V$

This is a positive number, so manganate(VII) ions, in acid solution, will oxidise chloride ions. (The reaction is thermodynamically spontaneous.)

Method 2

The value of $E^{\ominus}_{cell}$ can also be obtained by reversing the half-equation containing substance B and at the same time changing the sign of its $E^{\ominus}$ value. Then add it to the half-equation for substance A. This is the better method if the overall equation is also required.

> ### Worked example
>
> Will iron(III) ions oxidise iodide ions to iodine?
>
> $$Fe^{3+} + e^- \rightleftharpoons Fe^{2+} \qquad E^{\ominus} = +0.77\,V$$
> $$\tfrac{1}{2}I_2 + e^- \rightleftharpoons I^- \qquad E^{\ominus} = +0.54\,V$$
>
> *Answer*
>
> The reactants are Fe^{3+} (on the left of its half-equation) and I^- (on the right of its half-equation).
>
> The I_2/I^- half-equation is reversed, its sign changed and then it is added to the Fe^{3+}/Fe^{2+} half-equation.
>
> $$I^- \rightarrow \tfrac{1}{2}I_2 + e^- \qquad E^{\ominus} = -(+0.54)\,V = -0.54\,V$$
> $$Fe^{3+} + e^- \rightarrow Fe^{2+} \qquad E^{\ominus} = +0.77\,V$$
>
> Adding gives:
>
> $$Fe^{3+} + I^- \rightarrow Fe^{2+} + \tfrac{1}{2}I_2 \quad E^{\ominus}_{cell} = +0.77\,V + (-0.54\,V) = +0.23\,V$$
>
> This is a positive number, so the reaction is feasible and iron(III) ions oxidise iodide ions.

Writing overall redox equations

Questions often provide the necessary reduction half-equations and ask you to write an overall equation. This is done in four steps:

- **Step 1:** Look at the overall reaction and identify the *reactants* in the two half-equations. One reactant (the oxidising agent) is on the left-hand side of one half-equation and the other reactant (the reducing agent) is on the right-hand side of the other half-equation.

- **Step 2:** Reverse the equation that has the reactant on the *right*.

- **Step 3:** Multiply one or both equations, so that both now have the same number of electrons. One equation will still have the electrons on the left and the other will have them on the right.

- **Step 4:** Add the two equations. *Cancel* the electrons, as there will be the same number on each side of the equation.

content guidance

Worked example

Write the overall equation for the oxidation of ethanedioate ions ($C_2O_4^{2-}$) by manganate(VII) ions (MnO_4^-) in acid solution and predict whether it will be thermodynamically spontaneous. The two reduction half-equations are:

$$MnO_4^- + 8H^+ + 5e^- \rightleftharpoons Mn^{2+} + 4H_2O \qquad\qquad E^\ominus = +1.52\,V$$
$$2CO_2 + 2e^- \rightleftharpoons C_2O_4^{2-} \qquad\qquad E^\ominus = -0.49\,V$$

Answer

Step 1: the reactants are MnO_4^- and $C_2O_4^{2-}$

Step 2: reverse the second equation:

$$C_2O_4^{2-} \rightleftharpoons 2CO_2 + 2e^- \qquad\qquad E^\ominus = -(-0.49) = +0.49\,V$$

Note that when a half-equation is reversed, the sign of its $E^\ominus$ value is changed

Step 3: multiply the first equation by two and the reversed second equation by five so that both half-equations have the same number of electrons:

$$2MnO_4^- + 16H^+ + 10e^- \rightleftharpoons 2Mn^{2+} + 8H_2O \qquad\qquad E^\ominus = +1.52\,V$$
$$5C_2O_4^{2-} \rightleftharpoons 10CO_2 + 10e^- \qquad\qquad E^\ominus = +0.49\,V$$

Note that when a half-equation is multiplied by a number, its value is *not* altered.

Step 4: add to give the overall equation:

$$5C_2O_4^{2-} + 2MnO_4^- + 16H^+ \rightleftharpoons 10CO_2 + 2Mn^{2+} + 8H_2O$$

Tip Always check that your ionic equation balances for charge. In this example, both sides are 4+.

The value of $E^\ominus_{\text{reaction}}$ is found by adding the two $E^\ominus$ values at step 3:

$$E^\ominus_{\text{reaction}} = +1.52 + (+0.49) = +2.01\,V$$

This is a positive number and so the reaction is thermodynamically spontaneous (feasible).

When predictions and experiment do not agree

There could be several reasons for this.

Kinetic reasons

The activation energy of a reaction that is thermodynamically spontaneous (positive $E^\ominus_{\text{cell}}$) may not take place if the activation energy is too high.

An example of this is the oxidation of ethanedioate ions by manganate(VII) ions. The value of $E^\ominus_{\text{cell}}$ is +2.01 V, and so the reaction is classified as being thermodynamically spontaneous. But this reaction does not happen at room temperature, as the activation energy is too high. On warming, the reaction proceeds quickly enough for the two substances to be titrated.

Note that Mn^{2+} ions, which are one product, are a catalyst for this redox reaction.

Concentration effects: non-standard conditions

Aqueous copper(II) ions should not be reduced to aqueous copper(I) ions by iodide ions:

$$Cu^{2+}(aq) + e^- \rightleftharpoons Cu^+(aq) \qquad E^\ominus = +0.15\,V$$
$$I_2(aq) + 2e^- \rightleftharpoons 2I^-(aq) \qquad E^\ominus = +0.54\,V$$

Copper(II) ions are the oxidising agent and iodide ions the reducing agent, so for the reaction (equation I):

I $\quad Cu^{2+}(aq) + I^-(aq) \rightleftharpoons Cu^+(aq) + \frac{1}{2}I_2(aq) \;\; E^\ominus_{cell} = +0.15 - (+0.54) = -0.39\,V$

$E^\ominus_{cell}$ is negative, so this reaction is *not* thermodynamically spontaneous under standard conditions. However, when aqueous copper(II) and iodide ions are mixed, the iodide ions do reduce copper(II) ions to copper(I). The reason can be seen in the equation for the reaction that does happen (equation II):

II $\quad Cu^{2+}(aq) + 2I^-(aq) \rightarrow CuI(s) + \frac{1}{2}I_2(aq)$

Copper(I) iodide is totally insoluble and so $[Cu^+(aq)]$ is close to zero. This drives the equilibrium for equation I over to the right. The conditions are not standard, so that although $E^\ominus_{cell}$ is negative, E_{cell} for the *non-standard* conditions is positive.

Reaction of vanadium ions

Vanadium exists in several oxidation states:

Oxidation state	Formula of ion	Colour
+5	VO_3^-	Colourless
	VO_2^+	Yellow
+4	VO^{2+}	Blue
+3	V^{3+}	Green
+2	V^{2+}	Lavender

When vanadate(V) ions (VO_3^-) are acidified, an equilibrium is established between two vanadium(V) species.

$$VO_3^-(aq) + 2H^+(aq) \rightleftharpoons VO_2^+(aq) + H_2O(l)$$

Both vanadium species are in the +5 state, so this is *not* a redox reaction.

Vanadate(V) ions, in acid solution, can be reduced to different extents by different reducing agents.

The $E^{\ominus}_{cell}$ values for the different reductions are shown in the table below.

Oxidation state change	Equation	$E^{\ominus}$ /V
$5 \rightarrow 4$	$VO_2^+ + 2H^+ + e^- \rightleftharpoons VO^{2+} + H_2O$	$+1.00$
$4 \rightarrow 3$	$VO_2^+ + 2H^+ + e^- \rightleftharpoons V^{3+} + H_2O$	$+0.34$
$3 \rightarrow 2$	$V^{3+} + e^- \rightleftharpoons V^{2+}$	-0.26

- A reducing agent that appears on the *right* of a redox half-equation and which has a standard reduction potential value greater than $+1.0\,V$ will not reduce vanadium(v) ions.
- A reducing agent that appears on the *right* of a redox half-equation and which has a standard reduction potential value between $+0.34$ and $+1.0\,V$ will reduce vanadium(v) to the $+4$ state but no further.
- A reducing agent that appears on the *right* of a redox half-equation which has a standard reduction potential value between $-0.26\,V$ and $+0.34\,V$ will reduce vanadium(v) to the $+3$ state but not to $+2$.
- A reducing agent that appears on the *right* of a redox half-equation which has a standard reduction potential less than (more negative than) $-0.26\,V$ will reduce vanadium(v) $+5$ to the $+2$ state.

Worked example

How far will a solution of Fe^{2+} ions reduce a solution of VO_2^+?

$$Fe^{3+} + e^- \rightleftharpoons Fe^{2+} \qquad E^{\ominus} = +0.77\,V$$

Answer

The figure $+0.77\,V$ is between $+0.34\,V$ and $+1.0\,V$, and so Fe^{2+} should reduce vanadium(v) to vanadium(iv) and no further. To prove this, reverse the Fe^{3+}/Fe^{2+} equation and add it to the one for vanadium changing oxidation state from $+5$ to $+4$.

$$Fe^{2+} + VO_2^+ + 2H^+ \rightleftharpoons Fe^{3+} + VO^{2+} + H_2O$$
$$E^{\ominus}_{cell} = +1.0 + (-0.77) = +0.23\,V$$

As $E^{\ominus}_{cell}$ is positive, the reaction will occur. Therefore, Fe^{2+} will reduce vanadium $+5$ to the $+4$ state.

However, reversing the Fe^{3+}/Fe^{2+} equation and adding it to the one for vanadium $+4$ to $+3$ gives:

$$Fe^{2+} + VO^{2+} + 2H^+ \rightleftharpoons Fe^{3+} + V^{3+} + H_2O$$
$$E^{\ominus}_{cell} = +0.34 + (-0.77) = -0.43\,V$$

$E^{\ominus}_{cell}$ is negative, so the reaction will not occur. The Fe^{2+} ions will not reduce the solution to the $+3$ state.

This can be tested in the laboratory by adding excess iron(iii) sulfate solution to acidified ammonium vanadate(v). The solution goes from yellow to blue, showing that the vanadium is now in the $+4$ state.

Estimating the concentration of a solution of an oxidising agent

The general method is to add a known volume of a solution of the oxidising agent from a pipette to an *excess* of acidified potassium iodide and then titrate the liberated iodine with standard sodium thiosulfate solution. Starch is added when the iodine has faded to a pale straw colour, and the sodium thiosulfate solution is then added drop by drop until the blue-black starch–iodine colour disappears.

The equation for the titration reaction is:

$$I_2 + 2Na_2S_2O_3 \rightarrow 2NaI + Na_2S_4O_6$$

This is an equation that you must know and you must be able to use its stoichiometry to calculate the amount of iodine that reacted with the sodium thiosulfate in the titre. For example, if the iodine produced required 12.3 cm^3 of 0.456 mol dm^{-3} sodium thiosulfate solution, then:

amount (moles) of $Na_2S_2O_3$ = 0.456 mol dm^{-3} × 0.0123 dm^3 = 0.005609 mol

amount (moles) of I_2 = $\frac{1}{2}$ × 0.005609 = 0.00280 mol

 Tip

Remember that:
- moles of solute = concentration (in mol dm^{-3}) × volume (in dm^3)
- volume in dm^3 = $\dfrac{\text{volume in cm}^3}{1000}$

Worked example

Some 25.00 cm^3 of iron(III) chloride solution were added to excess acidified potassium iodide solution. The liberated iodine required 24.20 cm^3 of 0.100 mol dm^{-3} sodium thiosulfate solution to remove the colour. Calculate the concentration of the iron(III) chloride solution.

Answer

Redox equation: $2FeCl_3 + 2KI \rightarrow I_2 + 2FeCl_2 + 2KCl$

Titration equation: $I_2 + 2Na_2S_2O_3 \rightarrow 2NaI + Na_2S_4O_6$

Amount of thiosulfate = 0.100 mol dm^{-3} × $\dfrac{24.20}{1000 \text{ dm}^3}$ = 0.002420 mol

Amount of iodine = $\frac{1}{2}$ × 0.002420 = 0.001210 mol ($\frac{1}{2}$ because 1 mol I_2 reacts with 2 moles of $Na_2S_2O_3$)

Amount of $FeCl_3$ = 2 × 0.001210 = 0.002420 mol (2× because 2 moles of $FeCl_3$ produce 1 mol of I_2)

Concentration of the $FeCl_3$ solution = 0.002420 mol ÷ 0.02500 dm^3
= 0.0968 mol dm^{-3}

Estimating the concentration of reducing reagents

Almost all reducing agents reduce potassium manganate(VII) in acid solution. The colour of potassium manganate(VII) is so intense that there is no need to add an indicator.

- Transfer 25.00 cm³ of the solution of the reducing agent from a pipette into a conical flask.
- Acidify with approximately 25 cm³ of dilute sulfuric acid.
- Rinse a burette with water and then with some of a standard solution of potassium manganate(VII).
- Add the potassium manganate(VII) solution from the burette until a faint, permanent, pink colour is obtained.
- Repeat until two consistent titres are obtained and find the average of their values.

Worked example

'Iron' tablets contain iron(II) sulfate. The percentage of iron in these tablets can be found as follows:

- Weigh one iron tablet.
- Crush the tablet and dissolve it in about 25 cm³ of dilute sulfuric acid.
- Use distilled water to make the solution up to 250 cm³.
- Titrate 25.00 cm³ portions with standard 0.0202 mol dm⁻³ potassium manganate(VII) solution until two consistent titres have been obtained.

Data

Mass of tablet = 10.31 g

Mean titre = 26.20 cm³

Equation: $MnO_4^- + 8H^+ + 5Fe^{2+} \rightarrow Mn^{2+} + 4H_2O + 5Fe^{3+}$

Answer

Amount of MnO_4^- = 0.0202 mol dm⁻³ × (26.20 dm³ ÷ 1000)
                      = 0.0005292 mol

Amount of Fe^{2+} in 25.00 cm³ of solution = 5 × 0.0005292 = 0.002646 mol

Amount of Fe^{2+} in 250 cm³ of solution = 10 × 0.002646 = 0.02646 mol

Mass of iron in 1 tablet = moles × molar mass of Fe
                         = 0.02646 mol × 55.8 g mol⁻¹
                         = 1.476 g

Percentage of iron in one tablet = (1.476 × 100) ÷ 10.31 = 14.3%

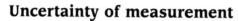

Uncertainty of measurement

All measurements have an in-built error, however careful the experimenter.

Titrations

Burette readings are only accurate to $\pm 0.05 \, \text{cm}^3$. As two readings are taken each titration, the combined error is $\pm 0.1 \, \text{cm}^3$.

If the titre is $20.00 \, \text{cm}^3$, the possible error due to the apparatus is $(0.1 \times 100) \div 20.00 = 0.5\%$.

For a titre of $10.00 \, \text{cm}^3$, the percentage error would be 1%.

Weighing

Chemical balances are more accurate and weigh to $\pm 0.01 \, \text{g}$, but masses of a substance require two readings, so the possible error is $0.02 \, \text{g}$. The percentage error in a weighing of a substance of mass $10.00 \, \text{g}$ is 0.2%.

Enthalpy experiments

Temperature-change measurements in enthalpy experiments are less accurate. Usually, a thermometer can be read to $\pm 0.2°C$, so the possible error in ΔT is $\pm 0.4°C$. For a temperature rise of $7.0°C$, the percentage error would be $(0.4 \times 100) \div 7.0 = 6\%$

Thus, an enthalpy of neutralisation experiment giving a value of $\Delta H^{\ominus} = -55 \, \text{kJ mol}^{-1}$ could be $\pm 55 \times \frac{6}{100} = \pm 3 \, \text{kJ}$. So the measured value can be stated as lying between -52 and $-58 \, \text{kJ mol}^{-1}$.

Worked example

A sample of mass $1.30 \, \text{g}$ of a d-block metal sulfate $M_2(SO_4)_3$ was weighed out and dissolved in water. Excess potassium iodide was added and the liberated iodine titrated against standard sodium thiosulfate solution. The titre showed that $0.00325 \, \text{mol}$ of the transition-metal sulfate was present in the $2.00 \, \text{g}$.

(a) Calculate the molar mass of the $M_2(SO_4)_3$.

(b) Hence, calculate the atomic mass of the element M. Suggest its identity.

(c) Assume that the error in each weighing is $\pm 0.01 \, \text{g}$, calculate the percentage error in the mass of $M_2(SO_4)_3$ taken.

(d) Assuming that there are no other significant errors, calculate the maximum and minimum values of the molar mass and hence of the atomic mass of the element M.

(e) What does this indicate about the reliability of the identification of M?

Answer

(a) molar mass $= \dfrac{\text{mass}}{\text{moles}} = \dfrac{1.30\,\text{g}}{0.00325\,\text{mol}} = 400\,\text{g mol}^{-1}$

(b) Of this 400, the three sulfate groups have a mass of
$3 \times (32.1 + (4 \times 16)) = 288.3$

Mass due to two atoms of M $= 400 - 288.3 = 111.7$

Relative atomic mass of M $= \frac{1}{2} \times 111.7 = 55.85$, and so the element M is iron, as the value in the periodic table is 55.8.

(c) The total weighing error is $\pm 0.02\,\text{g}$, so the percentage error in weighing the solid:

$= \frac{0.02}{130} \times 100 = 1.54\%$

(d) $\pm 1.54\%$ in $400\,\text{g mol}^{-1}$ is $\pm \frac{1.54}{100} \times 400 = \pm 6.2$ in the molar mass, so the maximum molar mass is $406.2\,\text{g mol}^{-1}$ and the minimum is $393.8\,\text{g mol}^{-1}$, and the maximum relative atomic mass is $\frac{1}{2} \times (406.2 - 288.3) = 58.95$ and the minimum relative atomic mass is $\frac{1}{2} \times (393.8 - 288.3) = 52.75$.

(e) These values suggest that the metal M could be manganese ($A_r = 54.9$), iron ($A_r = 55.8$) or chromium ($A_r = 58.9$), and so the validity of the identification as iron is poor.

Fuel cells

The internal combustion engine is less than 50% efficient in converting chemical energy into the kinetic energy of the car. The combustion of fossil fuels also produces greenhouse gases. Fuel cells may be the way forward in reducing greenhouse-gas emissions as the percentage conversion of chemical energy to kinetic energy is high.

The simplest cell, as used in space exploration, has hydrogen as the fuel, oxygen as the oxidant, and an electrolyte of aqueous potassium hydroxide. Each reacts on the surface of a catalyst. The hydrogen is oxidised at one electrode and releases electrons:

$$H_2(g) + 2OH^-(aq) \rightarrow 2H_2O(l) + 2e^- \qquad E^{\ominus} = +0.83\,\text{V}$$

These electrons pass round a circuit, powering an electric motor. When they arrive at the other electrode, they reduce oxygen:

$$\tfrac{1}{2}O_2(g) + H_2O(l) + 2e^- \rightarrow 2OH^-(aq) \qquad E^{\ominus} = +0.40\,\text{V}$$

The overall reaction is:

$$H_2(g) + \tfrac{1}{2}O_2(g) \rightarrow H_2O(l) \qquad E^{\ominus}_{\text{cell}} = +0.83 + (+0.40) = +1.23\,\text{V}$$

Such a fuel cell is not carbon neutral, as the hydrogen has to be manufactured either from methane or by the electrolysis of water. The generation of electricity produces carbon dioxide, either by burning fossil fuels or from the building of wind turbines or nuclear power stations.

Other fuels are being developed. Methanol and bioethanol are possibilities.

$$C_2H_5OH(aq) + 12OH^-(aq) \rightarrow 2CO_2(g) + 9H_2O(l) + 12e^-$$

$$3O_2(g) + 6H_2O(l) + 12e^- \rightarrow 12OH^-(aq)$$

$$C_2H_5OH(aq) + 3O_2(g) \rightarrow 2CO_2(g) + 3H_2O(l) \qquad E^{\ominus}_{cell} = +1.15\,V$$

Breathalysers

A breathalyser is a device used by police forces to measure the amount of alcohol in a driver's exhaled breath. The earliest device was a tube containing crystals of potassium dichromate(VI) and acid. The crystals are orange and are reduced to green Cr^{3+} ions by ethanol. The extent to which the tube goes green measures the amount of alcohol in the person's breath and hence in their bloodstream. This method is not accurate and the tubes cannot be used again.

Infrared spectrometers are also used to determine the amount of alcohol in a driver's breath. The machine measures the height of an absorption peak in the fingerprint region of ethanol. The drawback is that the peak height is not directly proportional to the amount of alcohol, so a doubling of the alcohol content produces an increase, but not a doubling, of the peak height.

Another device measures the total electrical output per given volume of breath. This is a fuel cell with the gaseous ethanol in the breath as the fuel and the oxygen in the exhaled breath as the oxidant. This is highly accurate, because the quantity of electricity is directly proportional to the amount of ethanol in a given volume of breath.

Transition-metal chemistry

Here are some important definitions:
- **d-block elements** are those in which the last electron has gone into a d-orbital.
- A **transition element** has partially filled d-orbitals in one or more of its ions. Transition elements are all d-block elements. All are metals; most are physically strong and have high melting points.
- A **ligand** is a molecule or negative ion that uses a lone pair of electrons to form a dative bond with a d-block cation.
- The **coordination number** is the number of atoms, ions or groups datively bonded to a d-block cation.
- A **dative (coordinate) bond** is formed when two atoms share a pair of electrons, both of which are supplied by one atom.

> **Tip** When defining a d-block element, do not say 'outermost electrons are in d-orbitals', or 'the highest energy electron is in a d-orbital', because neither is true.

Electron configuration

The electronic configurations of the d-block elements are shown in the table below.

Element	Electron configuration	3d	4s
Sc	[Ar] $3d^1\ 4s^2$	↑ ☐ ☐ ☐ ☐	↑↓
Ti	[Ar] $3d^2\ 4s^2$	↑ ↑ ☐ ☐ ☐	↑↓
V	[Ar] $3d^3\ 4s^2$	↑ ↑ ↑ ☐ ☐	↑↓
Cr	[Ar] $3d^5\ 4s^1$	↑ ↑ ↑ ↑ ↑	↑
Mn	[Ar] $3d^5\ 4s^2$	↑ ↑ ↑ ↑ ↑	↑↓
Fe	[Ar] $3d^6\ 4s^2$	↑↓ ↑ ↑ ↑ ↑	↑↓
Co	[Ar] $3d^7\ 4s^2$	↑↓ ↑↓ ↑ ↑ ↑	↑↓
Ni	[Ar] $3d^8\ 4s^2$	↑↓ ↑↓ ↑↓ ↑ ↑	↑↓
Cu	[Ar] $3d^{10}\ 4s^1$	↑↓ ↑↓ ↑↓ ↑↓ ↑↓	↑
Zn	[Ar] $3d^{10}\ 4s^2$	↑↓ ↑↓ ↑↓ ↑↓ ↑↓	↑↓

Note: [Ar] is short for the electronic configuration of argon, which is $1s^2\ 2s^2\ 2p^6\ 3s^2\ 3p^6$.

All the elements have the configuration [Ar] $3d^x\ 4s^2$, except chromium ([Ar] $3d^5\ 4s^1$) and copper ([Ar] $3d^{10}\ 4s^1$). The difference is caused by the extra stability of a half-filled or fully filled orbital type, which makes it energetically preferable for an electron to move out of the $4s$-orbital into a $3d$-orbital. An atom of iron has six $3d$-electrons because it is the sixth d-block element. The electronic configurations of atomic iron, the Fe^{2+} ion and the Fe^{3+} ion are:

Element	Electron configuration	3d	4s
Fe	[Ar] $3d^6\ 4s^2$	↑↓ ↑ ↑ ↑ ↑	↑↓
Fe^{2+}	[Ar] $3d^6$	↑↓ ↑ ↑ ↑ ↑	☐
Fe^{3+}	[Ar] $3d^6$	↑ ↑ ↑ ↑ ↑	☐

When a transition metal loses electrons and forms a cation, the first electrons lost come from the $4s$-orbital. Any further electrons come from paired electrons in d-orbitals. Thus the Fe^{3+} ion has no $4s$-electrons and five unpaired $3d$-electrons.

Tip All d-block cations have a $4s^0$ electronic configuration.

Evidence for electronic configurations

There is a significant jump in the successive ionisation energies between removing the last 4s-electron and the first 3d-electron. For chromium and copper, this jump is between the first and second ionisation energies. For the other d-block elements, it lies between the second and third ionisation energies.

There is also a big jump after the last 3d-electron has been removed, as the next electron comes from a 3p-orbital that is much more strongly held. For vanadium, this is after the fifth ionisation energy. For chromium, it is after the sixth; for manganese after the seventh; and for iron after the eighth ionisation energy. This is the evidence for chromium and copper having only one 4s-electron, and for the total number of 4s- and 3d-electrons.

Note that the second ionisation energy is the energy required to remove the second electron from the 1+ gaseous ion, and is not the energy required to remove two electrons from the gaseous atom.

Properties of transition elements

Variable oxidation state

Transition elements have several different oxidation states. The common oxidation states of the d-block metals are shown in the table below.

Sc	Ti	V	Cr	Mn	Fe	Co	Ni	Cu	Zn
								+1	
	+2	+2	+2	+2	+2	+2	+2	+2	+2
+3	+3	+3	+3		+3	+3			
	+4	+4		+4					
		+5							
			+6	+6					
				+7					

- Scandium forms only Sc^{3+} ions, which have no d-electrons; zinc forms only Zn^{2+} ions which have ten d-electrons. Therefore, neither is classified as a transition element.
- Chromium can be:
  - Cr^{2+}, e.g. $CrCl_2$
  - Cr^{3+}, e.g. $Cr_2(SO_4)_3$
  - in the +6 state, e.g. K_2CrO_4 and $K_2Cr_2O_7$
- Iron can be:
  - Fe^{2+}, e.g. $FeSO_4$
  - Fe^{3+}, e.g. $FeCl_3$
- Copper can be:
  - Cu^+, e.g. $CuCl$ and Cu_2O
  - Cu^{2+}, e.g. $CuSO_4$ and CuO

Transition metals can form stable cations with different charges because the successive ionisation energies increase steadily. For example, the extra energy required to remove a third electron from Fe^{2+} is compensated for by the *extra hydration enthalpy* of the 3+ ion compared with the 2+ ion. The third ionisation energy of iron ($Fe^{2+}(g) \rightarrow Fe^{3+}(g) + e^-$) is $+2960\,kJ\,mol^{-1}$, but the hydration energy of the $Fe^{3+}(g)$ ion ($Fe^{3+}(g) + aq \rightarrow Fe^{3+}(aq)$) is $2920\,kJ\,mol^{-1}$ *more exothermic* than the hydration energy of the $Fe^{2+}(g)$ ion.

With calcium, an s-block element, the third ionisation energy is $+4940\,kJ\,mol^{-1}$, which is not compensated for by the extra hydration enthalpy. The third ionisation energy is much higher because the third electron in calcium is removed from an inner $3p$-shell. A transition metal such as chromium can also form oxo-anions, such as CrO_4^{2-}, because it uses all its $4s$- and $3d$-electrons in forming six covalent bonds (4σ and 2π) with oxygen. It accepts six electrons to have the configuration $3d^{10}\,4s^2$.

Coloured complex ions

The six ligands around the central ion split the d-orbitals into a group of three of lower energy and a group of two of higher energy. When light is shone into a solution of a complex ion, the ion absorbs light energy and an electron is promoted to the upper of the two split d-levels. If both red light and yellow light are absorbed, the ion appears blue.

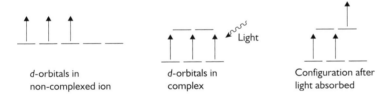

d-orbitals in non-complexed ion | d-orbitals in complex | Configuration after light absorbed

- Sc^{3+} and Ti^{4+} ions have no d-electrons. The Cu^+ and Zn^{2+} ions have a full set of 10 d-electrons. Therefore, no d–d transitions are possible. Hence, these ions are colourless.
- $[Cr(H_2O)_6]^{2+}$ is blue; $[Cr(H_2O)_6]^{3+}$ appears green.
- $[Fe(H_2O)_6]^{2+}$ is green; $[Fe(H_2O)_6]^{3+}$ is amethyst (solutions appear yellow-brown owing to deprotonation of the hydrated ion).
- $[Cu(H_2O)_6]^{2+}$ is pale blue; $[Cu(NH_3)_4(H_2O)_2]^{2+}$ is dark blue.

Note that anhydrous copper(II) sulfate is white. There are no ligands, so the d-electrons are not split, so no promotion is possible. This means that none of the white light is absorbed.

Formation of complex ions

All transition-metal ions form complexes. A complex ion is formed when a number of ligands (usually four or six) bond to a central metal ion. The bonding is dative covalent, with a lone pair of electrons on the ligand forming a bond with empty $3d$-, $4s$- and $4p$-orbitals in the metal ion.

Monodentate ligands

These use one pair of electrons to form a dative bond with the metal ion.

- Monodentate ligands can be neutral molecules, such as H_2O and NH_3, or negative ions, such as Cl^- and CN^-.
- All transition metals form hexaqua hydrated ions, such as $[Cr(H_2O)_6]^{3+}$.
- Zinc, a *d*-block metal, forms the $[Zn(H_2O)_4]^{2+}$ ion.
- In aqua complexes, the bonding between the oxygen atoms and the central metal ion is dative covalent, and the bonding within the water molecule is covalent.
- Fe^{2+} and Fe^{3+} ions form complexes with cyanide ions, $[Fe(CN)_6]^{4-}$ and $[Fe(CN)_6]^{3-}$.
- Cu^{2+} ions form a complex with ammonia — $[Cu(NH_3)_4(H_2O)_2]^{2+}$.
- Chloride ions are much larger than water molecules, so there is a maximum of four Cl^- ions around a transition-metal ion in chloro-complexes. For example, the complexes of copper(I) and copper(II) with chloride ions have the formulae $[CuCl_2]^-$ and $[CuCl_4]^{2-}$.

Bidentate ligands

These are molecules or ions that have two atoms with lone pairs which are both used in forming dative bonds with the metal ion. An example of a neutral bidentate ligand is 1,2-diaminoethane ($NH_2CH_2CH_2NH_2$). This used to be called ethylenediamine and was given the shorthand symbol 'en'. Three 1,2-diaminoethane molecules form six dative bonds with transition-metal ions such as Cu^{2+}:

$$[Cu(H_2O)_6]^{2+}(aq) + 3en(aq) \rightarrow [Cu(en)_3]^{2+}(aq) + 6H_2O(l)$$

The driving force for this ligand-exchange reaction is the gain in entropy as four particles turn into seven particles.

Note that the ring formed between the bidentate ligand and the metal ion must contain five or six atoms, otherwise there would be too much strain on the ring.

An example of a bidentate anion is the ethanedioate ion, $^-OOC-COO^-$. This used to be called the oxalate ion and had the symbol 'ox^{2-}'.

$$[Cu(H_2O)_6]^{2+}(aq) + 3ox^{2-}(aq) \rightarrow [Cu(ox)_3]^{4-}(aq) + 6H_2O(l)$$

Polydentate ligands

EDTA (the old name was ethylenediaminetetraacetic acid) is a 4^- ion that can form six ligands. The stability constants of these complexes are high:

$$[Cu(H_2O)_6]^{2+}(aq) + EDTA^{4-} \rightarrow [Cu(EDTA)]^{2-}(aq) + 6H_2O(l)$$

Haemoglobin contains an iron ion complexed with a pentadentate ligand occupying five sites, with an oxygen atom in the sixth.

Shapes of complex ions

All six-coordinate complexes are *octahedral* with 90° bond angles, because there are six pairs of bonding electrons around the metal ion and these six *bond pairs* repel each other to a position of *maximum separation*.

Tip Do not say that the atoms or the bonds repel. It is the bond pairs of electrons that repel.

If there are four ligands, the shape is either square-planar, as in the platinum(II) complex $[Pt(NH_3)_2(Cl_2)]$ or tetrahedral, as in the chromium(III) complex $[CrCl_4]^{3-}$. If there are only two ligands, the complex is linear, as in the dichlorocopper(I) anion, $[CuCl_2]^-$.

Catalytic activity

Transition metals and their compounds are often excellent catalysts. The metals use their *d*-orbitals to provide active sites on their surfaces to which reactants bond. For example, nickel is the catalyst for the addition of hydrogen to alkene. Iron is used in the Haber process, catalysing the reaction:

$$N_2 + 3H_2 \rightleftharpoons 2NH_3$$

Compounds of transition metals can change oxidation state and this is made use of in industrial processes. Vanadium(V) oxide is the catalyst in the manufacture of sulfuric acid and catalyses the reaction:

$$2SO_2 + O_2 \rightleftharpoons 2SO_3$$

The compound can do this because of the variable valency of vanadium. The mechanism is:

- **Step 1:** $2SO_2 + 2V_2O_5 \rightarrow 2SO_3 + 4VO_2$
- **Step 2:** $4VO_2 + O_2 \rightarrow 2V_2O_5$
- **Overall:** $2SO_2 + O_2 \rightarrow 2SO_3$

The Fe^{3+} ions catalyse the oxidation of iodide ions by persulfate ions ($S_2O_8^{2-}$). The mechanism is:

- **Step 1:** $2Fe^{3+}(aq) + 2I^-(aq) \rightarrow 2Fe^{2+}(aq) + I_2(s)$
- **Step 2:** $2Fe^{2+}(aq) + S_2O_8^{2-}(aq) \rightarrow 2Fe^{3+}(aq) + 2SO_4^{2-}(aq)$
- **Overall:** $2I^-(aq) + S_2O_8^{2-}(aq) \rightarrow I_2(s) + 2SO_4^{2-}(aq)$

Compounds of some group 4 and 5 transition metals such as rhodium and iridium are used in industry as catalysts. The reaction of carbon monoxide with methanol to produce ethanoic acid is an industrial process with a high atom economy. It is catalysed by complex rhodium or iridium compounds.

$$CO + CH_3OH \rightarrow CH_3COOH$$

Reactions of *d*-block metal compounds

Redox reactions

The reduction of vanadium(v) compounds is described on p. 21.

Chromium can exist in the +6 state, as in compounds such as potassium dichromate(VI) ($K_2Cr_2O_7$), which is orange and potassium chromate(VI) (K_2CrO_4), which is yellow. It also forms many compounds in the +3 state, such as chromium(III) sulfate ($Cr_2(SO_4)_3$) and many complexes of Cr^{3+} ions. Most chromium(III) compounds are green.

Some chromium(II) compounds also exist, but most of these are unstable. The chromium(II) complex with ethanoate ions is stable. It is an unusual complex, in that it contains two Cr^{2+} ions.

The standard reduction potentials are listed below:

$$\tfrac{1}{2}Cr_2O_7{}^{2-} + 7H^+ + 3e^- \rightleftharpoons Cr^{3+} + \tfrac{7}{2}H_2O \qquad E^\ominus = +1.33\,\text{V}$$

$$Cr^{3+} + e^- \rightleftharpoons Cr^{2+} \qquad\qquad\qquad\qquad E^\ominus = -0.41\,\text{V}$$

Reduction of chromium compounds

A reducing agent that appears on the *right* of a redox half-equation and which has a standard reduction potential value of less than +1.33 V will reduce dichromate(VI) ions in acid solution under standard conditions.

A reducing agent that appears on the *right* of a redox half-equation and which has a standard reduction potential value that is *more* negative than −0.41 V will reduce Cr^{3+} to Cr^{2+}.

Worked example 1

Will bromide ions reduce acidified dichromate ions to Cr^{3+} ions but no further under standard conditions?

$$\tfrac{1}{2}Br_2 + e^- \rightleftharpoons Br^- \qquad\qquad E^\ominus = +1.07\,\text{V}$$

Answer

For +6 to +3:

$$E^\ominus_{cell} = E^\ominus(Cr_2O_7{}^{2-}/Cr^{3+}) - E^\ominus(Br_2/Br^-) = +1.33 - (+1.07) = +0.26\,\text{V}$$

This is a positive number, so bromide ions will reduce acidified dichromate ions under standard conditions.

For +3 to +2:

$$E^\ominus_{cell} = E^\ominus(Cr^{3+}/Cr^{2+}) - E^\ominus(Br_2/Br^-) = -0.41 - (1.07) = -1.48\,\text{V}$$

This is a negative value, so bromide ions will not reduce Cr^{3+} ions to Cr^{2+} ions.

Worked example 2

Will zinc metal, in acid solution, reduce Cr^{3+} ions to Cr^{2+} ions?

$$Zn^{2+} + 2e^- \rightleftharpoons Zn \qquad\qquad E^{\ominus} = -0.76\,V$$

Answer

$$E^{\ominus}_{cell} = E^{\ominus}(Cr^{3+}/Cr^{2+}) - E^{\ominus}(Zn^{2+}/Zn) = -0.41 - (-0.76) = +0.35\,V$$

This is a positive number, so zinc metal will reduce chromium(III) ions to chromium(II) ions.

Oxidation of chromium compounds

Chromium(III) is oxidised to chromium(VI) in alkaline solutions:

$$CrO_4^{2-} + 4H_2O + 3e^- \rightleftharpoons Cr(OH)_3 + 5OH^- \qquad\qquad E^{\ominus} = -0.13\,V$$

An oxidising agent that appears on the *left* of a redox half-equation and which has a standard reduction potential which is more positive than $-0.13\,V$ will oxidise chromium(III) compounds to chromate(VI) in alkaline solution.

Worked example

Will hydrogen peroxide, H_2O_2, in alkaline solution, oxidise chromium(III) sulfate to chromate(VI) ions?

$$H_2O_2 + 2e^- \rightleftharpoons 2OH^- \qquad\qquad E^{\ominus} = +1.24\,V$$

Answer

$$E^{\ominus}_{cell} = E^{\ominus}(H_2O_2/OH^-) - E^{\ominus}(Cr^{3+}/CrO_4^{2-}) = +1.24 - (-0.13) = +1.37\,V$$

This is a positive number so hydrogen peroxide, in alkaline solution, will oxidise chromium(III) to chromate(VI).

Preparation of chromium(II) ethanoate complex

A solution of potassium dichromate(VI) is placed in a flask with some zinc and a mixture of 50% concentrated hydrochloric acid and water is added. A delivery tube is fitted through a screw seal with the bottom end below the surface of the acidified potassium dichromate(VI) solution and the other end below the surface of a solution of sodium ethanoate.

At first the seal at the top is left completely loose in order to let out the hydrogen. The orange solution turns green as Cr^{3+} ions are formed, and then blue as these are reduced to Cr^{2+} ions. At this stage, the cap of the seal is screwed shut and the pressure of hydrogen forces the solution containing Cr^{2+} ions out into the test tube.

The hydrated chromium(II) ions undergo ligand exchange, forming a precipitate of the red *chromium(II) ethanoate* complex.

$$2[Cr(H_2O)_6]^{2+}(aq) + 4CH_3COO^-(aq) \rightarrow [Cr_2(CH_3COO)_4(H_2O)_2](s) + 10H_2O(l)$$

Disproportionation

Disproportionation occurs when an *element* in a *single* species is simultaneously oxidised and reduced. Whether this type of reaction takes place depends on the $E^{\ominus}$ values of the two redox half-equations.

Worked example

Indicate whether the following is a disproportionation reaction and whether it will take place under standard conditions. Justify your answer.

$$3MnO_4^{2-} + 4H^+ \rightleftharpoons MnO_2 + 2MnO_4^- + 2H_2O$$

Data:

$$MnO_4^- + e^- \rightleftharpoons MnO_4^{2-} \quad E^{\ominus} = +0.56\,V$$

$$MnO_4^{2-} + 4H^+ + 2e^- \rightleftharpoons MnO_2 + 2H_2O \quad E^{\ominus} = +1.41\,V$$

Answer

It is a disproportionation reaction, because the manganese in the single species manganate(vi) is oxidised to manganate(vii) in MnO_4^- and reduced to manganese(iv) in MnO_2.

$$E^{\ominus}_{cell} = E^{\ominus}(MnO_4^{2-}/MnO_2) - E^{\ominus}(MnO_4^{2-}/MnO_4^-) = +1.41 - (+0.56)$$

$$= +0.85\,V$$

This is a positive number, so manganate(vi) ions disproportionate spontaneously.

Ligand exchange

These are reactions in which one ligand either totally or partially replaces the ligand in a complex. For example, when an EDTA solution is added to a solution of the hexaquachromium(iii) ion, the EDTA complex is formed in a ligand-exchange reaction:

$$[Cr(H_2O)_6]^{3+} + EDTA^{4-} \rightarrow [Cr(EDTA)]^- + 6H_2O$$

If cyanide ions are added to hexaqua iron(iii) ions, ligand exchange takes place:

$$[Fe(H_2O)_6]^{3+} + 6CN^- \rightarrow [Fe(CN)_6]^{3-} + 6H_2O$$

If excess ammonia solution is added to precipitates of the hydroxides of copper(ii), zinc, nickel, cobalt or silver, ligand exchange takes place, forming a solution of the ammonia complex of the *d*-block ion.

$$[Cu(H_2O)_4(OH)_2] + 4NH_3 \rightarrow [Cu(NH_3)_4(H_2O)_2]^{2+} + 2OH^- + 2H_2O$$

$$[Zn(H_2O)_2(OH)_2] + 4NH_3 \rightarrow [Zn(NH_3)_4]^{2+} + 2OH^- + 2H_2O$$

When concentrated hydrochloric acid is added to a solution of hydrated copper(ii) ions, the solution turns green as a ligand-exchange reaction takes place.

$$[Cu(H_2O)_6]^{2+} + 4Cl^- \rightarrow [CuCl_4]^{2-} + 6H_2O$$

Tip The colour of the copper compound depends on the ligand. This is the easiest way to see if ligand exchange has taken place.

The aqua complex of copper(I) disproportionates spontaneously into copper metal and hydrated copper(II) ions:

$$2[Cu(H_2O)_4]^+ \rightarrow Cu + [Cu(H_2O)_6]^{2+} + 2H_2O$$

However, some copper(I) complexes are stable.

If ammonia solution is added to solid copper(I) chloride, a colourless ammonia complex is formed.

$$CuCl + 2NH_3 \rightarrow [Cu(NH_3)_2]^+ + Cl^-$$

This solution turns blue as air slowly oxidises it to the copper(II) ammonia complex. If concentrated hydrochloric acid is added to solid copper(I) chloride, a colourless copper(I) chloride complex is formed:

$$CuCl + Cl^- \rightarrow [CuCl_2]^-$$

Note that these two copper(I) complexes are colourless, because the copper ion has a full *d*-shell. Although the *d*-orbitals are split in energy by the ligands, no promotion is possible, as all the *d*-orbitals are full.

Deprotonation

With water

All aqua ions of transition metals are to some extent deprotonated by water. The extent depends on the charge density of the metal ion. The greater the charge and the smaller the radius, the more the aqua ion is deprotonated.

Solutions of hydrated chromium(III) ions are acidic, owing to the production of H_3O^+ ions:

$$[Cr(H_2O)_6]^{3+}(aq) + H_2O(l) \rightleftharpoons [Cr(H_2O)_5OH]^{2+}(aq) + H_3O^+(aq)$$

With aqueous sodium hydroxide

All the hydrated ions are further deprotonated by $OH^-(aq)$ ions. A precipitate of the neutral hydrated hydroxide is formed:

$$[M(H_2O)_6]^{2+}(aq) + 2OH^-(aq) \rightarrow [M(H_2O)_4(OH)_2](s) + 2H_2O(l)$$

$$[M(H_2O)_6]^{3+}(aq) + 3OH^-(aq) \rightarrow [M(H_2O)_3(OH)_3](s) + 3H_2O(l)$$

where M stands for any *d*-block metal cation.

Note that the number of hydroxide ions on the left-hand side and the number of water molecules on the right-hand side of the equation are equal to the charge on the *d*-block metal ion.

The colour changes are shown in the table below.

Complex	Colour of complex	Hydroxide	Comment
$[Cr(H_2O)_6]^{3+}$	Green solution	Green precipitate	—
$[Mn(H_2O)_6]^{2+}$	Pale pink solution	Sandy precipitate that darkens in air	Manganese is oxidised to the +4 state
$[Fe(H_2O)_6]^{2+}$	Pale green solution	Green precipitate that goes brown in air	Iron is oxidised to the +3 state
$[Fe(H_2O)_6]^{3+}$	Yellow solution	Red-brown precipitate	—
$[Co(H_2O)_6]^{2+}$	Pink solution	Blue precipitate that goes red on standing	—
$[Ni(H_2O)_6]^{2+}$	Green solution	Pale green precipitate	—
$[Cu(H_2O)_6]^{2+}$	Blue solution	Blue precipitate	—
$[Zn(H_2O)_6]^{2+}$	Colourless solution	White precipitate	—

Reactions of amphoteric hydroxides with excess aqueous sodium hydroxide

On adding excess sodium hydroxide, amphoteric hydroxides deprotonate further — for example, chromium and zinc hydroxides:

$$[Cr(H_2O)_3(OH)_3] + 3OH^- \rightarrow [Cr(OH)_6]^{3-} + 3H_2O$$
dark green

$$[Zn(H_2O)_2(OH)_2] + 2OH^- \rightarrow [Zn(OH)_4]^{2-} + 2H_2O$$
colourless

Reactions with aqueous ammonia

Ammonia is a base and a ligand. When ammonia solution is added to a solution of a transition-metal salt, the hydrated metal ion is deprotonated, as with sodium hydroxide solution, giving the same coloured precipitates. For example:

$$[Fe(H_2O)_6]^{2+} + 2NH_3 \rightarrow [Fe(H_2O)_4(OH)_2] + 2NH_4^+$$

$$[Fe(H_2O)_6]^{3+} + 3NH_3 \rightarrow [Fe(H_2O)_3(OH)_3] + 3NH_4^+$$

However, excess ammonia results in **ligand exchange** with nickel, copper and zinc.

- $[Ni(H_2O)_4(OH)_2] + 4NH_3 \rightarrow [Ni(NH_3)_4(H_2O)_2]^{2+} + 2OH^- + 2H_2O$
  pale blue

- $[Cu(H_2O)_4(OH)_2] + 4NH_3 \rightarrow [Cu(NH_3)_4(H_2O)_2]^{2+} + 2OH^- + 2H_2O$
  dark blue

- $[Zn(H_2O)_2(OH)_2] + 4NH_3 \rightarrow [Zn(NH_3)_4]^{2+} + 2OH^- + 2H_2O$
  colourless

> **Tip** Some insoluble silver compounds react with excess ammonia solution to form a colourless silver(I) complex — see the test for halides.

Some uses of transition metals and their compounds

Many transition metals have important uses. Iron is a structural material and is also used as a catalyst in the Haber process. Copper is used as an electrical conductor and as a decorative metal in brass. When alloyed with iron, chromium is used to form stainless steel, which resists corrosion.

Copper(I) chloride is used in some photochromic sunglasses. The glass contains small amounts of colourless copper(I) chloride and silver chloride. When ultraviolet light (UV) shines on the glasses, the copper(I) chloride reduces the silver chloride to silver, which darkens the glass. So it not only removes harmful UV rays, but also reduces the light intensity. When no longer exposed to UV light, the reaction reverses and the glass becomes totally transparent:

$$CuCl + AgCl \rightleftharpoons CuCl_2 + Ag$$

Organic chemistry: arenes

Required AS chemistry

- An **addition reaction** occurs when two substances combine to form a single substance. For example:

$$H_2C=CH_2 + Br_2 \rightarrow CH_2BrCH_2Br$$

- A **substitution reaction** occurs when an atom or group in one compound is replaced by an atom or group from another substance. For example:

$$CH_3CH_2Cl + NaOH \rightarrow CH_3CH_2OH + NaCl$$

- An **elimination reaction** occurs when the components of a simple molecule are removed from an organic molecule and are *not* replaced by other atoms or groups. For instance, a hydrogen atom and a halogen, could be removed by an alkali, forming a C=C group. This happens with 2-bromopropane in the presence of hot potassium hydroxide in ethanol:

$$CH_3CHBrCH_3 + KOH \rightarrow H_2C=CHCH_3 + KBr + H_2O$$

- An **electrophile** is an atom, ion or group that, when forming a covalent bond, attacks an electron-rich site and accepts a pair of electrons from that site. Examples include HBr, Br_2, NO_2^+, CH_3^+ and CH_3C^+O.

- A **nucleophile** is an atom, ion or group that attacks a $\delta+$ atom, forming a covalent bond by donating a lone pair of electrons to that atom. Examples include H_2O, NH_3, OH^- and CN^-.

- A **free radical** is an atom or group with an unpaired electron. Examples include $Cl\bullet$ and $CH_3\bullet$.

A2 chemistry of arenes

Arenes are sometimes called **aromatic compounds**. They contain a benzene ring.

Structure of benzene

Benzene (C_6H_6) is a cyclic compound that has six carbon atoms in a hexagonal ring. Early theories suggested that there were alternate single and double bonds between the carbon atoms, but this did not fit with later experimental evidence. It was shown that all the carbon–carbon bonds are the same length and that the molecule is planar.

Two modern theories are used to explain the structure:

- The Kekulé version assumes that benzene is a **resonance hybrid** between the two structures:

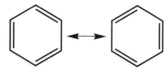

- The other theory assumes that each carbon atom is joined by a σ-bond to each of its two neighbours, and by a third σ-bond to a hydrogen atom. The fourth bonding electron is in a p-orbital, and the six p-orbitals overlap above and below the plane of the ring of carbon atoms. This produces a **delocalised** π-system of electrons, as in:

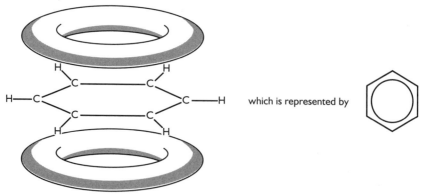

which is represented by

Thermochemical evidence: via enthalpy of hydrogenation

Benzene is more stable than 'cyclohexatriene', which is the theoretical compound with three single and three localised double carbon–carbon bonds. The amount by which it is stabilised can be calculated from the enthalpies of hydrogenation.

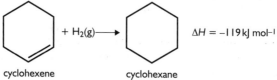

cyclohexene   + H_2(g) ⟶   cyclohexane $\Delta H = -119\,kJ\,mol^{-1}$

Therefore, ΔH for the addition to three localised double bonds in 'cyclohexatriene' would be $3 \times -119 = -357 \,\text{kJ}$. However:

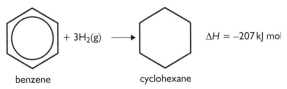

benzene   cyclohexane

Thus, 150 kJ *less* energy is given out because of benzene's unique structure. This is called the **delocalisation stabilisation energy** or **resonance energy** and can be shown in an enthalpy-level diagram.

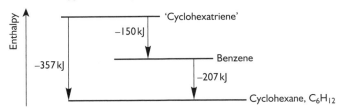

Thermochemical evidence: via bond enthalpies

The amount by which benzene is stabilised can also be calculated from average bond enthalpies. The enthalpy of formation of gaseous benzene is $+83 \,\text{kJ mol}^{-1}$.

The value for the theoretical molecule 'cyclohexatriene' can be found using the Hess's law cycle below:

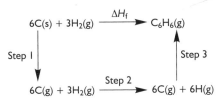

Step 1 equals $6 \times$ enthalpy of atomisation of carbon $(\Delta H_a) = 6 \times (+715) = +4290 \,\text{kJ}$

Step 2 equals $3 \times$ H—H bond enthalpy $= 3 \times (+436) = +1308 \,\text{kJ}$

Step 3 equals enthalpy change of bonds made:

- three C—C $= 3 \times (-348) = -1044 \,\text{kJ}$
- three C=C $= 3 \times (-612) = -1836 \,\text{kJ}$
- six C—H $= 6 \times (-412) = -2472 \,\text{kJ}$

Total $= -5352 \,\text{kJ}$

The ΔH_f of 'cyclohexatriene' $= \Delta H_{\text{step 1}} + \Delta H_{\text{step 2}} + \Delta H_{\text{step 3}} = +4290 + 1308 + (-5352)$ $= +246 \,\text{kJ mol}^{-1}$. This is 163 kJ more than the experimental value of 'real' benzene, which is $+83 \,\text{kJ mol}^{-1}$. The difference is the value of the resonance energy and is not quite the same as that calculated by the first method because it has been calculated using *average* bond enthalpies.

X-ray diffraction evidence

X-ray diffraction shows the position of the centre of atoms. If the diffraction pattern of benzene is analysed, it clearly shows that all the bond lengths between the carbon atoms are the same.

Bond	Bond length/nm
All six carbon–carbon bonds in benzene	0.14
Carbon–carbon single bond in cyclohexene	0.15
Carbon–carbon double bond in cyclohexene	0.13

Infrared evidence

Comparison of the infrared spectrum of aromatic compounds with those of aliphatic compounds containing a C=C group showed slight differences. The C—H stretching vibration in benzene is at $3036 \, cm^{-1}$ and the C=C stretching is at $1479 \, cm^{-1}$, whereas the equivalent vibrations in an aliphatic compound such as cyclohexene are at 3023 and $1438 \, cm^{-1}$.

Reactions of benzene

Combustion

Benzene burns in a limited amount of air with a smoky flame. Combustion is incomplete and particles of carbon are formed.

Addition

The double bond in benzene is not as susceptible to addition as is the double bond in alkenes. However, it does react with hydrogen in the presence of a hot nickel catalyst to form cyclohexane.

$$C_6H_6 + 3H_2 \rightarrow C_6H_{12}$$

Electrophilic substitution

A catalyst is *always* needed, since the stability due to delocalisation causes the reactions to have large activation energies.

Reaction with bromine: halogenation

Dry benzene reacts with liquid bromine in the presence of iron (or a catalyst of anhydrous iron(III) bromide). Steamy fumes of hydrogen bromide are given off and bromobenzene (C_6H_5Br) is formed.

The mechanism for this reaction is as follows. The catalyst, anhydrous iron(III) bromide, is made by the reaction of iron with bromine.

$$Fe + \tfrac{3}{2}Br_2 \rightarrow FeBr_3$$

This then reacts with more bromine, forming the electrophile Br$^+$:

$$Br_2 + FeBr_3 \rightarrow Br^+ + [FeBr_4]^-$$

The Br$^+$ attacks the π-electrons in the benzene ring, forming an intermediate with a positive charge. Finally, the [FeBr$_4$]$^-$ ion removes an H$^+$ from benzene, producing hydrogen bromide (HBr) and re-forming the catalyst (FeBr$_3$).

The addition of Br$^+$ to benzene is similar to the first step of the addition of bromine to ethene. The difference arises at the next step. The benzene intermediate loses an H$^+$, thus regaining *the stability of the delocalised π-system*, whereas the intermediate with ethene adds a Br$^-$ ion. A catalyst must be present for the addition of Br$^+$ to benzene, because the activation energy of the first step is higher than that for the addition to ethene.

Reaction with nitric acid: nitration

When benzene is warmed with a mixture of concentrated nitric and sulfuric acids, a nitro-group (NO$_2$) replaces a hydrogen atom in the benzene ring. Nitrobenzene and water are produced.

The sulfuric acid reacts with the nitric acid to form the electrophile NO$_2$$^+$. The temperature must not go above 50°C or some dinitrobenzene (C$_6$H$_4$(NO$_2$)$_2$) is formed.

Tip The mechanisms of bromination and nitration could be shown as two separate steps.

Reaction with fuming sulfuric acid: sulfonation

When benzene is warmed with fuming sulfuric acid, benzenesulfonic acid is produced. Fuming sulfuric acid is a solution of sulfur trioxide in sulfuric acid. The electrophile is the SO$_3$ molecule.

This reaction is important in the manufacture of detergents, where a substituted benzene ring is sulfonated and the product neutralised.

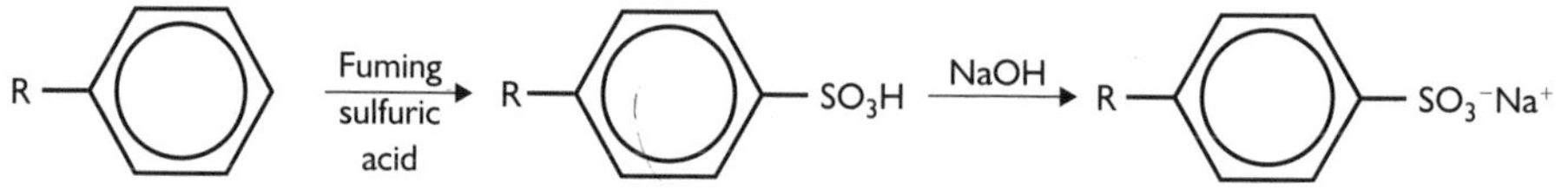

R represents a long hydrophilic hydrocarbon chain.

Friedel–Crafts reaction

Reaction with halogenoalkanes

In the presence of an anhydrous aluminium chloride catalyst, alkyl groups (e.g. C_2H_5) can be substituted into the ring. In the reaction between benzene and iodoethane, the products are ethylbenzene and hydrogen iodide.

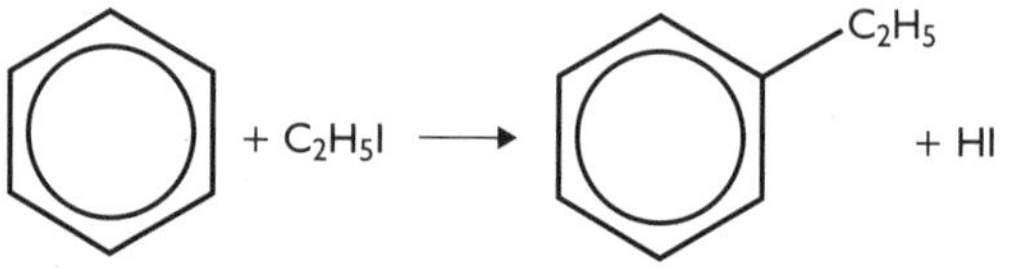

The reaction mixture must be *dry*.

The electrophile is produced by the reaction of the catalyst with the halogenoalkane:

$$CH_3CH_2I + AlCl_3 \rightarrow CH_3CH_2^+ + [AlCl_3I]^-$$

The positive carbon atom attacks the π-system in the benzene ring:

Step I:

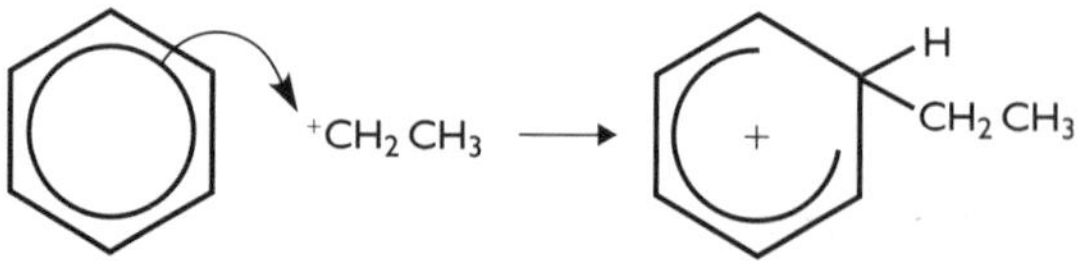

Tip Make sure that you draw the curly arrow towards the carbon atom in the CH_2 group and not the CH_3 group.

The intermediate loses a H^+ ion so as to *regain* the stability of the benzene ring.

Step 2:

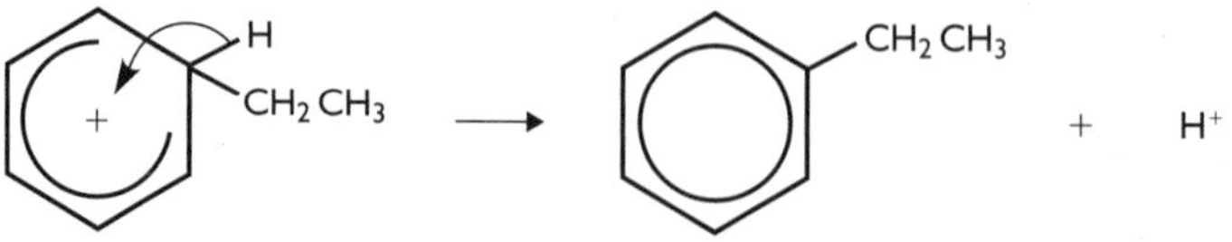

The catalyst is regenerated by the reaction:

$$H^+ + [AlCl_3I]^- \rightarrow HI + AlCl_3$$

Reaction with acid chlorides

In the presence of an anhydrous aluminium chloride catalyst, benzene reacts with acid chlorides to form ketones. In the reaction between benzene and ethanoyl chloride, the products are phenylethanone and hydrogen chloride.

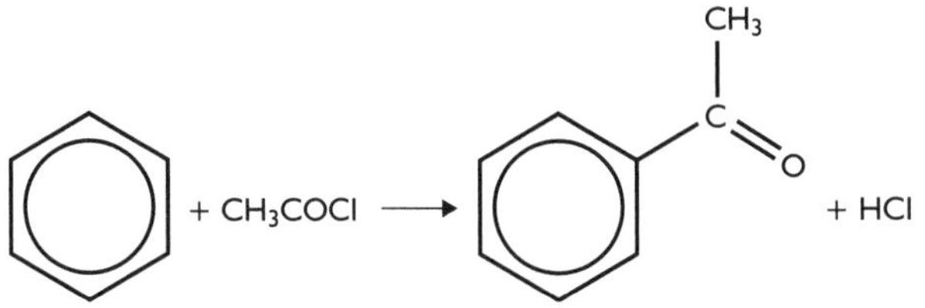

$$+ CH_3COCl \longrightarrow \qquad + HCl$$

The reagents must be *dry*.

The electrophile is produced by the reaction of the acid chloride with the catalyst:

$$CH_3COCl + AlCl_3 \rightarrow CH_3C^+O + [AlCl_4]^-$$

The positive carbon atom attacks the π-system in the benzene ring.

Step 1:

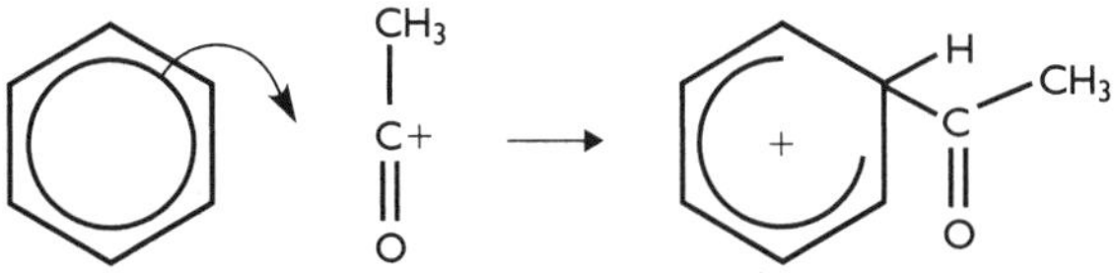

The intermediate loses a H^+ ion so as to *regain* the stability of the benzene ring.

Step 2:

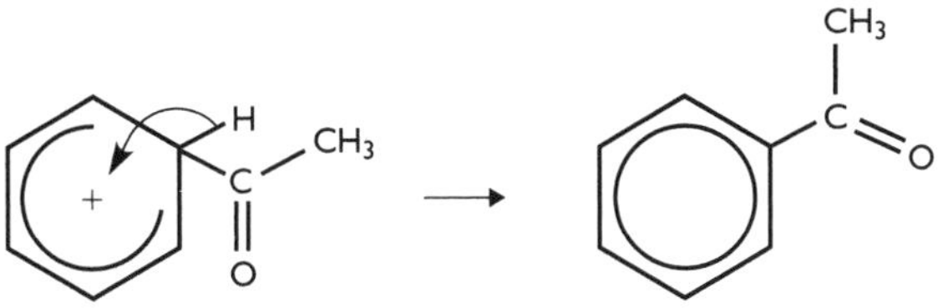

The catalyst is regenerated by the reaction:

$$H^+ + [AlCl_4]^- \rightarrow HCl + AlCl_3$$

Tips for benzene mechanisms
Make sure that:
- the curly arrow starts from the delocalised ring
- the intermediate has a delocalised ring around five carbon atoms and is positively charged
- the curly arrow starts from the σ-bond between a carbon and the hydrogen in the final step

Phenol

Phenol (C_6H_5OH) contains an —OH group on a benzene ring. A lone pair of electrons on the oxygen atom becomes part of the delocalised π-system and makes phenol more susceptible to attack by electrophiles.

Electrophilic substitution

Reaction with bromine
The electron-rich ring in phenol is attacked by bromine water, in an electrophilic substitution reaction. The brown bromine water is decolorised and a white precipitate

of 2,4,6-tribromophenol and a solution of hydrogen bromide are formed. No catalyst is needed.

Reaction with nitric acid

The ring is sufficiently activated for nitration to take place with dilute nitric acid. At room temperature, the organic product is a mixture of 2-nitrophenol and 4-nitrophenol.

If the mixture is heated, 2,4- and 2-6-dinitrophenol are formed as well. If concentrated nitric acid is used, 2,4,6-trinitrophenol is the product.

Organic chemistry: nitrogen compounds

The types of organic nitrogen compounds that are studied in Unit 5 are:

- **amines**: these contain a nitrogen atom covalently bonded to a carbon atom, usually in a NH_2 group, as in 1-butylamine ($CH_3CH_2CH_2CH_2NH_2$)
- **amides**: these contain the $-C\underset{NH_2}{\overset{O}{\Vert}}$ group
- **nitriles**: these contain the $C{\equiv}N$ group, as in propanenitrile (CH_3CH_2CN)
  Note: this has the propan- stem, as there are three carbon atoms in the chain.
- **condensation** polymers
- **amino acids**: these contain a NH_2 and a COOH group, as in aminoethanoic acid (NH_2CH_2COOH), which is also called glycine
- **diazo** compounds: these contain the $-N^+{\equiv}N$ ion, but are only formed from aromatic amines

Amines

Primary amines contain the $-NH_2$ group — for example, ethylamine ($C_2H_5NH_2$).

Secondary amines contain the $-NH$ group — for example, diethylamine ($(C_2H_5)_2NH$).

Tertiary amines contain the $-N<$ group — for example, triethylamine ($(C_2H_5)_3N$). They all contain a lone pair of electrons on the nitrogen atom, which is $\delta-$.

Physical properties

- Amines have a fish-like smell.
- Primary and secondary amines form intermolecular hydrogen bonds between the lone pair of electrons on the nitrogen in one molecule and the $\delta+$ hydrogen in another molecule.
- All amines form hydrogen bonds with water. The lone pair of electrons on the $\delta-$ nitrogen forms a hydrogen bond with the $\delta+$ hydrogen in water. This means that the members of the homologous series of amines of low molecular weight are miscible with water. Others, such as phenylamine ($C_6H_5NH_2$) are partially soluble.

Reactions

With water

Amines are weak bases and so deprotonate water in a reversible reaction:

$$C_2H_5NH_2 + H_2O \rightleftharpoons C_2H_5NH_3^+ + OH^-$$

As OH^- ions are formed, the solution is alkaline (pH > 7).

Note: Compare this reaction to that of ammonia and water:

$$NH_3 + H_2O \rightleftharpoons NH_4^+ + OH^-$$

With acids

Amines are weak bases, just like ammonia, and so they react with both strong and weak acids to form salts. With hydrochloric acid, the product is ethylammonium chloride:

$$C_2H_5NH_2 + HCl \rightarrow C_2H_5NH_3^+Cl^-$$

This is similar to the reaction of ammonia with hydrochloric acid:

$$NH_3 + HCl \rightarrow NH_4^+Cl^-$$

With a weak acid, such as ethanoic acid, the product is ethylammonium ethanoate:

$$C_2H_5NH_2 + CH_3COOH \rightarrow C_2H_5NH_3^+CH_3COO^-$$

With acid chlorides

Amines react rapidly with acid chlorides to form a substituted amide:

$$C_2H_5NH_2 + CH_3COCl \rightarrow CH_3CONHC_2H_5 + HCl$$

The organic product is called *N*-ethylethanamide. The *N*-ethyl part of the name shows that the ethyl group is attached to the *nitrogen* atom in the amide.

With copper(II) ions

Aliphatic amines form complex ions similar to that formed by ammonia:

$$[Cu(H_2O)_6]^{2+} + 4C_2H_5NH_2 \rightarrow [Cu(C_2H_5NH_2)_4(H_2O)_2]^{2+} + 4H_2O$$

$$[Cu(H_2O)_6]^{2+} + 4NH_3 \rightarrow [Cu(NH_3)_4(H_2O)_2]^{2+} + 4H_2O$$

The ethylamine reacts with hydrated copper(II) ions in a ligand-exchange reaction to form a blue solution.

With halogenoalkanes

When a primary amine is warmed with a halogenoalkane in ethanolic solution, the salt of the secondary amine is formed:

$$C_2H_5NH_2 + C_2H_5Cl \rightarrow (C_2H_5)_2NH_2{}^+Cl^-$$

The secondary amine can be freed by adding aqueous alkali and warming:

$$(C_2H_5)_2NH_2{}^+Cl^- + OH^- \rightarrow (C_2H_5)_2NH + H_2O + Cl^-$$

Phenylamine

Phenylamine ($C_6H_5NH_2$) is an arene and an amine.

Preparation from nitrobenzene

If nitrobenzene is heated under reflux with tin and concentrated hydrochloric acid, the NO_2 group is reduced to an $NH_3{}^+Cl^-$ group. Phenylamine is set free on addition of alkali. The overall reaction is:

$$C_6H_5NO_2 + 6[H] \rightarrow C_6H_5NH_2 + 2H_2O$$

Physical properties

Phenylamine is a volatile liquid which is partially soluble in water. The $\delta+$ hydrogen atoms form hydrogen bonds with the $\delta-$ oxygen atoms of the water, and the $\delta-$ nitrogen bonds with the $\delta+$ hydrogen atoms from the water.

When an acid is added, a solution is formed. This is because phenylamine is a base and forms an ionic salt with acids.

Reactions of the NH$_2$ group

As a base

The NH_2 group is a much weaker base than ethylamine. This is because the lone pair of electrons on the nitrogen atom becomes partially incorporated into the delocalised ring of the C_6H_5 group.

With water, the result is an alkaline solution plus some undissolved phenylamine:

$$C_6H_5NH_2(l) + H_2O(l) \rightleftharpoons C_6H_5NH_3{}^+(aq) + OH^-(aq)$$

With dilute hydrochloric acid, a solution of phenylammonium chloride is formed:

$$C_6H_5NH_2(l) + HCl(aq) \rightarrow C_6H_5NH_3{}^+(aq) + Cl^-(aq)$$

With ethanoyl chloride

Phenylamine reacts in the same way as aliphatic primary amines, to form an amide:

$$C_6H_5NH_2 + CH_3COCl \rightarrow CH_3CONHC_6H_5 + HCl$$

With copper(II) ions

Ligand exchange takes place, and a green solution of the phenylamine copper(II) complex is formed. Aromatic amines react in a similar manner to aliphatic amines, such as ethylamine, but the stoichiometry is different. There is not room for four bulky phenylamine molecules around the copper ion:

$$[Cu(H_2O)_6]^{2+} + 2C_6H_5NH_2 \rightarrow [Cu(C_6H_5NH_2)_2(H_2O)_4]^{2+} + 2H_2O$$

With nitrous acid

If a solution of phenylamine in dilute hydrochloric acid is added to a solution of sodium nitrite ($NaNO_2$) and the temperature is maintained between 0 and 10°C, a solution of benzenediazonium ions is formed:

$$C_6H_5NH_2 + NO_2^- + 2H^+ \rightarrow C_6H_5N^+\equiv N + 2H_2O$$

If the temperature drops below 0°C, then the reaction is too slow. If it is allowed to rise above 10°C, the diazonium compound decomposes. Benzenediazonium ions are too unstable to keep, but they react with phenol to form a yellow precipitate of a dye.

Reactions of the benzene ring

The ring is activated in the same way as with phenol. Thus, it reacts with bromine water to form 2,4,6-tribromophenylamine and hydrogen bromide.

Amides

Amides are prepared by the reaction of an acid chloride with ammonia:

$$CH_3COCl + NH_3 \rightarrow CH_3CONH_2 + HCl$$
ethanoyl chloride   ethanamide

Amides are hydrolysed by heating with dilute hydrochloric acid:

$$CH_3CONH_2 + HCl + H_2O \rightarrow CH_3COOH + NH_4Cl$$

Nitriles

Nitriles contain the $-C\equiv N$ group. The stem of their names is determined by the total number of carbon atoms in the chain, *including* the carbon of the CN group.

Reactions

Hydrolysis

When heated under reflux with an acid, a carboxylic acid is formed:

$$RCN + H^+ + 2H_2O \rightarrow RCOOH + NH_4^+$$

Hydroxynitriles, such as $CH_3CH(OH)CN$, also undergo this reaction. The organic product is 2-hydroxypropanoic acid (lactic acid):

$$CH_3CH(OH)CN + H^+ + 2H_2O \rightarrow CH_3CH(OH)COOH + NH_4^+$$

A nitrile can also be hydrolysed by heating under reflux with aqueous alkali:

$$RCN + OH^- + H_2O \rightarrow RCOO^- + NH_3$$

The salt of a carboxylic acid is produced. This can be converted to the carboxylic acid if dilute sulfuric acid is now added:

$$RCOO^- + H^+ \rightarrow RCOOH$$

Condensation polymers

Condensation polymers are formed when two compounds, each with two functional groups, polymerise. Water, or another simple molecule, is eliminated every time two molecules combine.

Polyamides

These are formed from a monomer containing two NH_2 groups and another containing two carboxylic acid or acid chloride groups.

Nylon is a polymer of 1,6-diaminohexane ($H_2N(CH_2)_6NH_2$) and hexane-1,6-dioyl chloride ($ClOC(CH_2)_4COCl$) with the elimination of HCl. The structure showing two repeat units is:

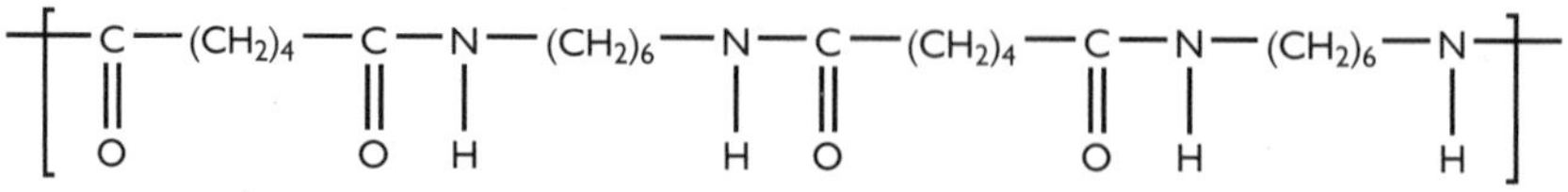

A polyamide

Kevlar is a polymer of 1,4-diaminobenzene ($H_2NC_6H_4NH_2$) and benzene-1,4-dicarbonyl dichloride ($ClOCC_6H_4COCl$).

One repeat unit is shown below.

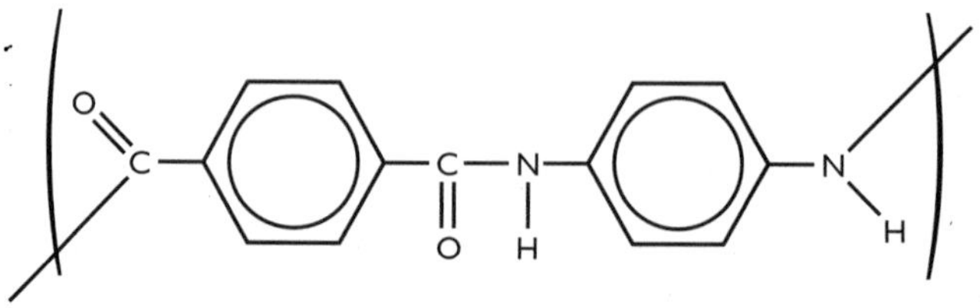

Tip Note that the repeat unit in all polyamides has two oxygen and two nitrogen atoms. Polyamides have a higher melting point than most polymers. This is because of hydrogen bonding between the $\delta+$ hydrogen of the NH group in one chain with the $\delta-$ oxygen in the C=O group of another chain.

Polyesters

These are another type of elimination polymer and are formed from one monomer containing two alcohol groups and another monomer with two acid or acid chloride groups.

One example is terylene. Ethane-1,2-diol ($HOCH_2CH_2OH$) reacts with benzene-1,4-dicarboxylic acid (also known as terephthalic acid, $HOOCC_6H_4COOH$) eliminating water and forming the polyester.

The repeat unit of this polymer is shown below:

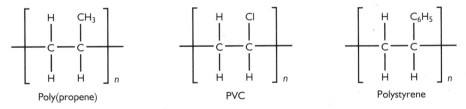

A polyester

Tip Note that the repeat unit has four oxygen atoms.

Addition polymers

Addition polymers are formed when compounds containing a C=C group undergo polymerisation. Examples include:
- ethene to poly(ethene)
- propene to poly(propene)
- chloroethene to poly(chloroethene), which is called PVC
- tetrafluoroethene to poly(tetrafluoroethene) or PTFE
- phenylethene to poly(phenylethene) or polystyrene

The repeat units of some of these polymers are shown below:

| Poly(propene) | PVC | Polystyrene |

Tip Note that all addition polymers have a carbon chain of two carbon atoms only. (The exception is if a 1,3-diene is polymerised.)

Some addition polymers have unusual properties when added to water.

| Propenamide | Poly(propenamide) |

Poly(propenamide) and poly(propenoic acid), especially when cross-linked, absorb water through hydrogen bonding. They are used in disposable nappies and as water-holding gels in the soil of pot plants and hanging baskets. As they absorb water they swell to many times their own volume.

Poly(ethenol)

This polymer is also called (polyvinyl alcohol). Its repeat unit is:

Tip The prefix 'vinyl' is sometimes used as the name for a $CH_2=CH$ group.

This polymer has the unusual property of being water-soluble. It is used as the coating in liquitabs, which contain liquid washing detergent, and in soluble laundry bags to hold soiled hospital laundry. The polymer is water-soluble because of the many hydrogen bonds that can form between the OH groups on every other carbon atom and water molecules.

Poly(ethenol) is unusual in another respect, in that its monomer does not exist. Any attempt to make ethenol, $CH_2=CHOH$, results in the production of ethanal, CH_3CHO. The polymer is manufactured by reacting the polyester, polyvinyl acetate (PVA), with methanol in a transesterification reaction.

Amino acids

Amino acids contain both $-NH_2$ and $-COOH$ groups. They are found naturally in proteins. Most contain a chiral centre. For example, 2-aminopropanoic acid ($CH_3CH(NH_2)COOH$) has two optical isomers.

Mirror

Physical properties

Effect on plane-polarised light

Most amino acids are chiral. Natural amino acids are single enantiomers of formula $H_2NCHRCOOH$. Natural alanine ($H_2NCH(CH_3)COOH$) is chiral and rotates the plane of polarisation of plane-polarised light. Glycine (H_2NCH_2COOH) is not chiral and so has no effect on plane-polarised light.

Solubility

Amino acids are water-soluble solids. The reason for this is that the acidic $-COOH$ group protonates the basic $-NH_2$ group, forming a **zwitterion**, which has a positive charge on one end and a negative charge on the other:

$$NH_2CH_2COOH \rightleftharpoons {}^+NH_3CH_2COO^-$$

The ion–dipole attractions involving the $\delta+$ H and $\delta-$ O atoms in the water give rise to its solubility.

Melting point

The ion–ion attractions between the *different* zwitterions result in the substance being a solid with a high melting point.

Reactions

With acids

The $-NH_2$ group becomes protonated:

$$NH_2CH_2COOH + H^+ \rightarrow {}^+NH_3CH_2COOH$$

With bases

The $-COOH$ protonates the base:

$$NH_2CH_2COOH + OH^- \rightarrow NH_2CH_2COO^- + H_2O$$

With ninhydrin

Ninhydrin reacts on heating with amino acids to form a deep-blue/purple colour. This reaction is used to identify the positions of different amino acids after chromatographic separation — see below.

Proteins

Proteins contain a sequence of amino acids joined by a peptide bond. This is the same bond that joins the monomers in a polyamide. If two amino acids join, two isomeric dipeptides are possible. When glycine and alanine react, the two possible dipeptides are:

The peptide link $-CONH-$ is circled in these two structures.

Proteins are polypeptides. Insulin has 17 different amino acids and a total of 51 amino acid molecules joined by 50 peptide bonds. This is an example of **condensation polymerisation**. The peptide links can be hydrolysed by prolonged heating with hydrochloric acid. This breaks the polypeptide down into its constituent amino acids, which can be separated by chromatography.

Thin-layer chromatography

In thin-layer chromatography (TLC), as with all types of chromatography, there is a stationary phase and a moving phase. In TLC, the stationary phase is either silica gel or aluminium oxide immobilised on a flat inert sheet, which is usually made from glass or plastic.

- The mixture of amino acids is dissolved in a suitable solvent and a spot of the solution placed about 2 cm from the bottom of the plate.
- Spots of dissolved known amino acids are placed on the same plate at the same level.
- The plate is then dipped in a suitable eluent (the mobile phase) with the spots above the level of the liquid eluent and is placed in a sealed container. The eluent is drawn up the plate by capillary action.
- The plate is left until the eluent rises to the top of the plate.
- The plate is removed, sprayed with a solution of ninhydrin and heated.
- The ninhydrin reacts with the amino acids, producing a blue-purple colour.
- The height that the unknown has reached is compared with the heights reached by the known amino acids. Spots at the same height are caused by the same amino acid. This enables the amino acids in the unknown to be identified.

Organic chemistry: analysis and synthesis

Analysis

Deduction of the molecular formula

The first step is finding the percentage composition of the substance. This is done by burning a known mass of the substance in excess air. The water is absorbed in a suitable drying agent, such as silica gel, and the carbon dioxide is absorbed by dry calcium oxide. This measures the masses of water and carbon dioxide formed. The route is then:

- mass of water → mass of hydrogen → % hydrogen
- mass of carbon dioxide → mass of carbon → % carbon
- 100 − (% of hydrogen + % of carbon) = % oxygen

Worked example

An organic compound X contains carbon, hydrogen and oxygen only. When 2.00 g of X was burnt in excess air, 1.46 g of water and 3.57 g of carbon dioxide were produced. Calculate the percentage composition of the elements in compound X.

Answer

1.46 g of H_2O contains $1.46 \times \frac{2}{18} = 0.162$ g hydrogen

% hydrogen $= 0.162 \times \frac{100}{2.00} = 8.1\%$

3.57 g of CO_2 contains $3.57 \times \frac{12}{44} = 0.9736$ g carbon

% carbon $= 0.9736 \times \frac{100}{2.00} = 48.7\%$

% oxygen $= 100 - 8.1 - 48.7 = 43.2\%$

The second step is to deduce the empirical formula from the percentage composition. The route for this is as follows:
- Divide each percentage by the relative atomic mass of the element.
- Divide by the smallest and round to one decimal place.
- If the answer is not a whole number ratio, multiply all by two (and if this still is not, try multiplying by three).

Worked example

Use your results from the Worked example above to calculate the empirical formula of compound X.

Answer

Element	%	% ÷ Ar	÷ smallest	× 2
Carbon	48.7	$\frac{48.7}{12} = 4.06$	$\frac{40.6}{2.7} = 1.5$	3
Hydrogen	8.1	$\frac{8.1}{1} = 8.1$	$\frac{8.1}{2.7} = 3.0$	6
Oxygen	43.2	$\frac{43.2}{16} = 2.7$	$\frac{2.7}{2.7} = 1.0$	2

The empirical formula of compound X is $C_3H_6O_2$.

The final step is to work out the molecular formula. For this, the relative molecular mass must be known. This will either be given or deduced from the mass spectrum, where the largest *m/e* value in the spectrum is assumed to be that of the molecular ion.

Worked example

The mass spectrum of compound X has a peak due to the molecular ion at $m/e = 74$. Use your results from the Worked example above to deduce its molecular formula.

Answer

The relative molecular mass is the same as the m/e of the molecular ion = 74.

The 'empirical mass' of $C_3H_6O_2$ = (3 × 12) + (6 × 1) + (2 × 16) = 74.

This is the same as the relative molecular mass, so the molecular formula is $C_3H_6O_2$.

Functional-group analysis

Chemical tests

You must give the *full* name or formula of the reagents used in any test. If a colour change is observed, then you must state the colour *before* and *after* the test.

Alkenes

The functional group is C=C. Add bromine dissolved in water. The brown solution becomes colourless.

Halogenoalkanes

Warm with a little aqueous sodium hydroxide mixed with ethanol. Then acidify with dilute nitric acid and add silver nitrate solution.

> **Tip** Make sure that you know the order in which the chemicals are added: sodium hydroxide, then nitric acid, then silver nitrate and finally ammonia.

The observation depends on the type of halogenoalkane:
- Chloroalkanes give a white precipitate soluble in dilute ammonia solution.
- Bromoalkanes give a cream precipitate insoluble in dilute ammonia solution but soluble in concentrated ammonia.
- Iodoalkanes give a yellow precipitate, insoluble in concentrated ammonia.

Hydroxyl groups in acids, alcohols and phenols

Add phosphorus pentachloride to the dry compound. Steamy fumes are given off.

Acids

Add aqueous sodium hydrogencarbonate. A gas is evolved that turns limewater cloudy.

Alcohols

Add ethanoic acid and a few drops of concentrated sulfuric acid and warm gently. Pour into a beaker of cold water and carefully smell the product. A fruity or glue-like smell confirms an alcohol.

To distinguish primary and secondary alcohols from tertiary alcohols, warm with potassium dichromate(VI) in dilute sulfuric acid:

- Primary and secondary alcohols turn the orange solution green.
- Tertiary alcohols do not react.

To distinguish between primary and secondary alcohols, test as above and distil the product into ammoniacal silver nitrate solution.

- Primary alcohols form a silver mirror.
- Secondary alcohols do not react.

Phenols
Add bromine water. The brown solution becomes colourless and a white precipitate forms.

Carbonyl compounds
Add 2,4-dinitrophenylhydrazine solution (Brady's reagent). A red-orange precipitate is produced.

To distinguish between aldehydes and ketones, add Tollens' reagent and warm.

- Aldehydes form a silver mirror.
- Ketones do not react.

Alternatively, add Fehling's (or Benedict's) solution and warm:

- Aldehydes produce a brown precipitate.
- Ketones do not react, so the blue solution remains.

Iodoform reaction
Add the test compound to a mixture of iodine and aqueous sodium hydroxide. A pale-yellow precipitate of iodoform, CHI_3, is produced with the following:

- methyl ketones, because they contain the CH_3CO group
- ethanal (CH_3CHO) and ethanol (CH_3CH_2OH)
- secondary alcohols, if they contain the $CH_3CH(OH)$ group, which is oxidised by the iodine to a methyl ketone

Physical methods
The infrared spectrum can give information about functional groups that are present in the molecule. The main frequencies to look for are:

- a peak at around $1700 \, cm^{-1}$. This denotes the presence of a $C=O$ group.
- a peak around $3100 \, cm^{-1}$. This denotes an OH group.

A more detailed list of bonds and frequencies can be found in a data book or textbook.

Identification of the compound
The molecular formula and the nature of the functional groups present may be enough to determine the identity of a substance. If compound X in the above Worked examples was shown to be a carboxylic acid, the only possible structure is CH_3CH_2COOH, so compound X is propanoic acid.

This would not be sufficient for an acidic compound of molecular formula $C_4H_8O_2$. It could be butanoic acid ($CH_3CH_2CH_2COOH$) or methylpropanoic acid (($CH_3)_2CHCOOH$). There are several ways in which the identity could be confirmed.

NMR spectral analysis
Butanoic acid has four peaks in its NMR spectrum. One is due to the hydrogen atoms in the CH_3 group, one to those in the CH_2 group next to the CH_3, one to those in the CH_2 group next to the COOH group, and one is due to the hydrogen in the OH group. The hydrogen atoms in the two CH_3 groups in methylpropanoic acid are in identical environments. Therefore, methylpropanoic acid has only three peaks.

The splitting pattern in an NMR spectrum can also be used to elucidate a structure. For example, the symmetrical isomers $CH_2OHCH=CHCH_2OH$ and $H_2C=C(CH_2OH)_2$ both have positive tests for alkenes and for primary alcohols. They both have three peaks in the NMR spectra, but the splitting patterns are different:
- The pattern for $CH_2OHCH=CHCH_2OH$ is a singlet (the OH peak), a doublet (the CH_2 peak) and a triplet (the CH peak).
- The pattern for $H_2C=C(CH_2OH)_2$ is three singlets, as OH is never split and the CH_2 hydrogen atoms do not have any hydrogen atoms on the neighbouring carbon atoms.

Melting- or boiling-point determination
The melting point of a solid or the boiling point of a liquid may be enough to identify it, if its molecular formula and functional groups are known. The melting/boiling point is determined (see p. 62) and its value compared with a source of data.

Preparation of a derivative, its purification and determination of its melting point
Boiling points are hard to measure accurately, so they may not be enough to identify a compound without doubt.

Both pentan-3-one ($CH_3CH_2COCH_2CH_3$) and pentan-2-one ($CH_3COCH_2CH_2CH_3$) have boiling points of 102°C. The compound 3-methylbutan-2-one ($CH_3COCH(CH_3)_2$) has a boiling point of 95°C.

They can be distinguished by reacting them with 2,4-dinitrophenylhydrazine. The solid formed is purified by recrystallisation (see p. 61) and its melting point is determined.

The derivative of pentan-3-one melts at 156°C, that of pentan-2-one at 144°C; and that of 3-methylbutanone at 123°C.

Synthesis

You are expected to be able to deduce methods of converting one organic substance into another. Such questions may involve a reaction scheme of up to four steps. You are required to identify:
- the reagent (by name or formula) for each step
- the conditions (if asked for)
- each intermediate in the synthesis

content guidance

Common reactions are listed below. Those marked '*' cause the carbon chain to be lengthened; that marked '#' causes the carbon chain to be shortened.

- *alkene* to a *halogenoalkane* — mix with gaseous hydrogen halide, e.g. hydrogen bromide, HBr
- *alkene* to a *halogenoalcohol* — bubble into bromine water
- *halogenoalkane* to an *alcohol* — heat under reflux with aqueous sodium hydroxide
- *halogenoalkane* to an *amine* — react with concentrated ammonia in aqueous ethanol
- *halogenoalkane* to an *alkene*— heat under reflux with potassium hydroxide in ethanol
- *alcohol* to a *halogenoalkane* — for example, add phosphorus pentachloride
- *primary alcohol* to an *aldehyde* (partial oxidation) — heat with potassium dichromate(VI) in dilute sulfuric acid and distil off the aldehyde as it forms
- *primary alcohol* to a *carboxylic acid* (complete oxidation) — heat under reflux with potassium dichromate(VI) in dilute sulfuric acid
- *secondary alcohol* to a *ketone* (oxidation) — heat under reflux with potassium dichromate(VI) in dilute sulfuric acid
- **aldehyde* or *ketone* to a *hydroxynitrile* — add hydrogen cyanide by reacting the carbonyl compound buffered at pH 6
- *nitrile* (or a hydroxynitrile) to a *carboxylic acid* (hydrolysis) — heat under reflux with dilute sulfuric acid
- *nitrile* to a *primary amine* (reduction) — lithium aluminium hydride in dry ether
- **benzene* to an *alkyl benzene* or a *ketone* (Friedel–Crafts reaction) — with a halogenoalkane or an acyl chloride and an anhydrous aluminium chloride catalyst
- *#methyl ketone* (e.g. $RCOCH_3$) or a *secondary 2-ol* (e.g. $CH_3CH(OH)R$) to a *carboxylic acid* (e.g. RCOOH) (iodoform reaction) — add iodine mixed with aqueous sodium hydroxide

Planning a synthesis

It is assumed that the reactions of compounds with more than one functional group are the same as those of simple compounds each with one functional group. For example, the reactions of $CH_2{=}CHCH_2CH(OH)CHO$ are those of an alkene, a secondary alcohol and an aldehyde.

Sometimes the conversion may require the carbon chain length to be altered, and this can be an important clue to the route. This means that one step in the reaction is the addition of hydrogen cyanide to a carbonyl compound, or a Friedel–Crafts reaction with an arene. If the length of the carbon chain is decreased, the iodoform reaction will be part of the synthesis.

Note that the carbon chain is the number of carbon atoms joined to each other. It is *not* necessarily the total number of carbon atoms in the molecule. For instance, the ester ethyl propanoate ($CH_3CH_2COOC_2H_5$) has two carbon chains, one with three carbon atoms (propanoate) and one with two (ethyl).

If you cannot immediately see the route, work backwards from the final product. For example, if the question asks how ethanoic acid can be prepared from ethane, ask yourself the following questions:

- How can an acid be prepared?

  Answer: by oxidation of a primary alcohol

- How can a primary alcohol be prepared?

  Answer: by hydrolysis of a halogenoalkane.

- Can a halogenoalkane be prepared from an alkene?

  Answer: yes, by addition of a hydrogen halide. The route is:

ethene $\xrightarrow{\text{HBr(g)}}$ bromoethane $\xrightarrow{\text{hot NaOH(aq)}}$ ethanol $\xrightarrow{\text{hot H}^+/\text{Cr}_2\text{O}_7^{2-}}$ ethanoic acid

Minimising risk in an experiment

Hazard data may be given in a question. This can be used to design an experiment which minimises the risks due to any hazards.

If a reactant or product:

- is poisonous or is a harmful or irritating gas, carry out the reaction in a fume cupboard
- is corrosive or is absorbed through the skin, wear disposable gloves
- is flammable, heat using a water-bath or an electric heater. This is essential if ether is present as the solvent of the reaction mixture, as in the reduction using lithium aluminium hydride.

Stereospecific reactions

These are particularly important in the pharmaceutical industry, as only one stereoisomer may be active as a medicine.

S_N1 reactions always give a racemic mixture, but an S_N2 reaction on a single enantiomer produces a single enantiomer.

The addition of bromine to an alkene proceeds via *trans*-addition, where one bromine atom adds on to one side of the double bond and the other to the opposite side. Thus, the addition to $CH_3CH=CHC_2H_5$ gives one optical isomer, whereas, if the addition were *cis*, a different optical isomer would be produced.

Experimental techniques

You should be able select suitable techniques for carrying out the reactions listed on p. 57. You should also be able to draw the sets of apparatus shown on the following page.

Heating under reflux

Heating under reflux is used when the reaction is slow and one of the reactants is volatile.

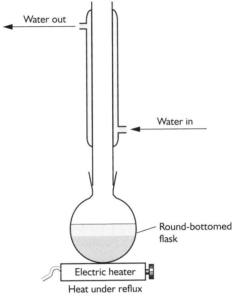

Heat under reflux

Distillation

This is used to remove a volatile substance from a mixture, e.g. a carboxylic acid from sulfuric acid and chromium compounds resulting from the oxidation of a primary alcohol.

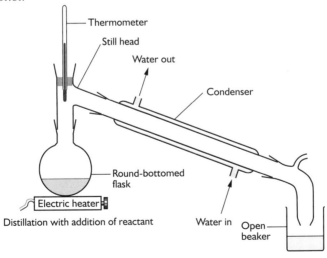

Distillation with addition of reactant

Steam distillation

This is used to extract a volatile substance that is insoluble in water from a reaction mixture that contains solid as well as a solution, e.g. phenylamine after reduction of nitrobenzene with tin and concentrated hydrochloric acid. It can also be used to extract volatile oils from natural products such as rose petals or orange peel.

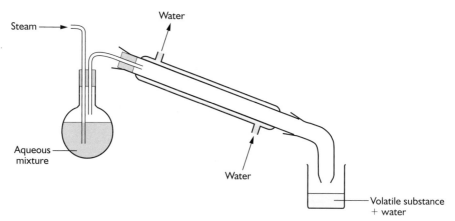

- In these diagrams it is always safer to draw an electric heater rather than a Bunsen burner, as a reactant or product might be flammable.
- Make sure that each piece of the apparatus is drawn as a separate piece and not fused to the next one. The joints between the different parts are difficult to draw.
- You must be able to draw a round-bottomed flask, a condenser (either in the reflux or distillation position) and a thermometer.
- Make sure that the water goes *in* at the *bottom* of the condenser and *out* at the *top*.
- Make sure that the apparatus is open to the air at some point — at the top of a reflux condenser or into a collecting vessel for distillation.

Solvent extraction

An organic product can often be separated from inorganic substances by solvent extraction. The compound 1-chloropentane can be prepared by adding excess phosphorus pentachloride to pentan-1-ol.

$$CH_3(CH_2)_3CH_2OH(l) + PCl_5(s) \rightarrow CH_3(CH_2)_3CH_2Cl(l) + HCl(g) + POCl_3(l)$$

Phosphorus oxychloride ($POCl_3$) has a boiling temperature of 105°C, which is similar to that of 1-chloropentane at 108°C, so it would not be separated by distillation of the reaction mixture.

The organic compound is separated by adding ether (ethoxyethane). This solvent dissolves the 1-chloropentane but not the phosphorus oxychloride. The ether solution is purified and the ether removed by distillation.

Purification techniques

At the end of a synthesis, the desired product is impure because of the presence of unused reactants or the products of side reactions, so it needs to be purified.

Purification of a solid

The solid is filtered off from the reaction mixture and purified by recrystallisation.

Recrystallisation
- Select a solvent in which the solid is soluble when hot and almost insoluble at room temperature.
- Dissolve the solid in the minimum amount of hot solvent.
- Filter the solution through a pre-heated glass filter funnel fitted with a fluted filter paper — this removes any insoluble impurities.
- Allow the filtrate to cool and crystals of the pure solid to appear.
- Filter using a Buchner funnel under reduced pressure — this removes any soluble impurities.
- Wash the solid with a little cold solvent and allow to dry.

Purification of a liquid that is insoluble in water
The product is distilled out of the reaction mixture and the following process is carried out.
- The distillate is washed with sodium carbonate solution in a separating funnel. The pressure must be released from time to time to let out the carbon dioxide. This washing is repeated until no more gas is produced. This removes any acidic impurities that may be present.

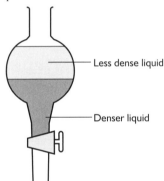

Less dense liquid

Denser liquid

- The aqueous layer is discarded and the organic layer washed with water. This removes any unreacted sodium salts and any soluble organic substances, such as ethanol.

> **Tip** The denser liquid will form the lower layer. You may assume that a mixture of organic substances will form a single layer.

- The aqueous layer is discarded and the organic layer dried, usually with lumps of anhydrous calcium chloride.
  Note: solid potassium hydroxide is used to dry amines and alcohols, as they form complex ions with calcium chloride.
- The organic liquid is decanted off from the drying agent and distilled, collecting the fraction that boils at ±2°C of the data-book boiling temperature.

Melting- and boiling-point determinations

Melting- and boiling-point determinations are used to identify a substance. The melting or boiling point is determined and checked against a data book.

Melting point

Heat the solid in a test tube immersed in a water-bath until the solid begins to melt. Measure the temperature at this point. When the substance has completely melted, allow it to cool, and then measure the temperature at which it begins to solidify. The substance must be stirred all the time. Average the two temperatures.

Boiling point

The method is as follows.

- Place a small amount of the test liquid in an ignition tube and, using a rubber band, attach it to the thermometer.
- Place an empty capillary tube in the liquid, with its open end below the surface.
- Clamp the thermometer in the beaker of water.
- Slowly heat the water, stirring all the time. When the stream of bubbles coming out of the capillary tube is rapid and continuous, note the temperature of the water bath and stop the heating process.
- Allow the beaker of water to cool, stirring continuously. Note the temperature of the water bath when bubbles stop coming out of the capillary tube and the liquid begins to suck back into the capillary tube.

The average of these two temperatures is the boiling point of the liquid. The apparatus used is shown below.

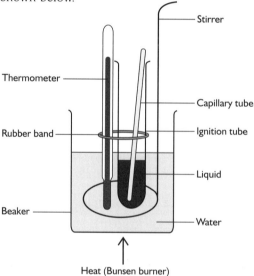

Combinatorial chemistry

The technique involves synthesis on the surface of resin beads a few micrometres in diameter. This can be illustrated by imagining the formation of a library of polyesters made up of two hydroxyacids of formula $HOCH_2CHXCOOH$, symbolised by the letters A and B. X represents a group such as H, CH_3 or C_2H_5 and the resin bead is symbolised by the letter R.

- **Step 1**: The two hydroxyacids are mixed with the resin and two products are obtained: R—A   R—B

- **Step 2**: The mix is then separated into two portions, each containing both of the products of the first step. Hydroxyacid A is added to one portion, hydroxyacid B to the second.

$$R{-}A \;\; R{-}B \qquad\qquad R{-}A \;\; R{-}B$$
$$\downarrow + A \qquad\qquad\qquad \downarrow + B$$
$$R{-}A{-}A \;\; R{-}B{-}A \qquad R{-}A{-}B \;\; R{-}B{-}B$$

This gives four diesters.

- **Step 3**: The diesters are mixed and split into two portions, each containing all four diesters. Hydroxyacid A is added to one portion and hydroxyacid B to the other. The addition of A to the first portion results in the production of four different triesters:

$$R{-}A{-}A{-}A, \;\; R{-}B{-}A{-}A, \; R{-}A{-}B{-}A \;\text{ and }\; R{-}B{-}B{-}A$$

The addition of B to the second portion results in four other triesters:

$$R{-}A{-}A{-}B, \;\; R{-}B{-}A{-}B, \; R{-}A{-}B{-}B, \;\text{ and }\; R{-}B{-}B{-}B.$$

Carrying out this process n times gives a library of $2n$ different polyesters.

Organic reactions: summary

Unit Test 5 is synoptic, in that questions may be asked on the organic chemistry in all four units.

Aliphatic chemistry

Alkanes

Reagent	Equation	Conditions
Halogens (e.g. chlorine)	$CH_4 + Cl_2 \rightarrow CH_3Cl + HCl$	Light (visible or UV)
Oxygen (combustion)	$2C_2H_6 + 7O_2 \rightarrow 4CO_2 + 6H_2O$	Burn or spark

Alkenes

In the table below, propene is used as an example.

Reagent	Equation	Conditions
Hydrogen	$CH_3CH{=}CH_2 + H_2 \rightarrow CH_3CH_2CH_3$	Heated nickel catalyst
Halogens (e.g. bromine)	$CH_3CH{=}CH_2 + Br_2 \rightarrow$ $CH_3CHBrCH_2Br$	Bubble propene into bromine dissolved in hexane
Aqueous halogens	$CH_3CH{=}CH_2 + Br_2 + H_2O \rightarrow$ $CH_3CH(OH)CH_2Br + HBr$	Bubble propene into bromine water
Hydrogen halides (e.g. hydrogen bromide)	$CH_3CH{=}CH_2 + HBr \rightarrow$ $CH_3CHBrCH_3$	Mix gases at room temperature
Potassium manganate(VII) (oxidation)	$CH_3CH{=}CH_2 + [O] + H_2O \rightarrow$ $CH_3CH(OH)CH_2OH$	Solution made alkaline with aqueous sodium hydroxide

Alkenes polymerise. For example, at 2000 atm and 250°C, propene polymerises to poly(propene):

$$nCH_3CH{=}CH_2 \rightarrow \text{+}CH(CH_3){-}CH_2\text{+}_n$$

Alcohols

Oxidation

Primary alcohols (e.g. propan-1-ol) are oxidised to carboxylic acids by heating under reflux with sulfuric acid and potassium dichromate(VI).

$$C_2H_5CH_2OH + 2[O] \rightarrow C_2H_5COOH + H_2O$$

However, if the oxidising agent is added to the hot alcohol and the product is distilled off as it is formed, propanal is produced.

$$C_2H_5CH_2OH + [O] \rightarrow C_2H_5CHO + H_2O$$

Secondary alcohols (e.g. propan-2-ol) can be oxidised to ketones by heating under reflux with sulfuric acid and potassium dichromate(VI).

$$CH_3CH(OH)CH_3 + [O] \rightarrow CH_3COCH_3 + H_2O$$

Tertiary alcohols (e.g. 2-methylpropan-2-ol) cannot be oxidised.

Other reactions

Other reactions of alcohols are summarised in the table below, using propan-1-ol as the example.

Reagent	Equation	Conditions
Sodium	$C_2H_5CH_2OH + Na$ $\rightarrow C_2H_5CH_2ONa + \frac{1}{2}H_2$	Add sodium metal to the alcohol
Phosphorus pentachloride	$C_2H_5CH_2OH + PCl_5 \rightarrow$ $C_2H_5CH_2Cl + HCl + POCl_3$	Dry the reagents at room temperature
Ethanoic acid	$C_2H_5CH_2OH + CH_3COOH$ $\rightarrow CH_3COOCH_2C_2H_5 + H_2O$	Warm gently with a few drops of concentrated sulfuric acid
Ethanoyl chloride	$C_2H_5CH_2OH + CH_3COCl$ $\rightarrow CH_3COOCH_2C_2H_5 + HCl$	Mix the dry reagents
Hydrogen halides, e.g. HBr	$C_2H_5CH_2OH + HBr$ $\rightarrow C_2H_5CH_2Br + H_2O$	HBr is made from 50% sulfuric acid and potassium bromide
or	or	
HI	$C_2H_5CH_2OH + HI$ $\rightarrow C_2H_5CH_2I + H_2O$	HI is made from damp red phosphorus and iodine

Halogenoalkanes

In the table below, 2-iodopropane is used as an example.

Reagent	Equation	Conditions
Aqueous sodium hydroxide	$CH_3CHICH_3 + NaOH$ $\rightarrow CH_3CH(OH)CH_3 + NaI$	Heat under reflux in aqueous solution
Water + silver nitrate solution Then:	$CH_3CHICH_3 + H_2O$ $\rightarrow CH_3CH(OH)CH_3 + H^+ + I^-$ $Ag^+ + I^- \rightarrow AgI(s)$	Warm gently Observe precipitate of silver halide
Ethanolic potassium hydroxide	$CH_3CHICH_3 + KOH$ $\rightarrow CH_3CH=CH_2 + KI + H_2O$	Heat under reflux in ethanolic solution
Ammonia	$CH_3CHICH_3 + 2NH_3$ $\rightarrow CH_3CH(NH_2)CH_3 + NH_4I$	Heat an ethanolic solution of ammonia in a sealed tube

Carbonyl compounds

Aldehydes and ketones

Ethanal is used as an example of an aldehyde and propanone as an example of a ketone.

- **Reaction with 2,4-dinitrophenylhydrazine**:

$CH_3CHO + H_2NNHC_6H_3(NO_2)_2 \rightarrow CH_3CH=NNHC_6H_3(NO_2)_2 + H_2O$

$(CH_3)_2C=O + H_2NNHC_6H_3(NO_2)_2 \rightarrow (CH_3)_2C=NNHC_6H_3(NO_2)_2 + H_2O$

The product is an orange precipitate.

- **Reaction with hydrogen cyanide**:

$CH_3CHO + HCN \rightarrow CH_3CH(OH)CN$

$CH_3COCH_3 + HCN \rightarrow CH_3C(OH)(CN)CH_3$

There must be both CN^- ions and HCN present.

- **Reduction with lithium aluminium hydride**:

$CH_3CHO + 2[H] \rightarrow CH_3CH_2OH$

$CH_3COCH_3 + 2[H] \rightarrow CH_3CH(OH)CH_3$

The reaction is carried out in a dry ether solution, *followed by* the addition of dilute acid.

Aldehydes

Aldehydes (e.g. ethanal) can be oxidised by warming gently with either Tollens' reagent or Fehling's (or Benedict's) solution.

$CH_3CHO + [O] + OH^- \rightarrow CH_3COO^- + H_2O$

Ethanoate ions are produced. Ammoniacal silver nitrate gives a silver mirror; Fehling's (or Benedict's) solution gives a red precipitate of copper(I) oxide.

Carbonyl compounds containing the CH_3CO group

If ethanal or methylketones are allowed to stand with iodine and sodium hydroxide, the following reaction occurs:

$$RCOCH_3 + 3I_2 + 4NaOH \rightarrow CHI_3 + 3NaI + RCOONa + 3H_2O$$

The products are a yellow precipitate of iodoform (CHI_3) and the sodium salt of a carboxylic acid that has one carbon atom fewer than the original carbonyl compound.

Carboxylic acids

In the table below, ethanoic acid is used as an example.

Reagent	Equation	Conditions
Alkalis (e.g. sodium hydroxide)	$CH_3COOH + NaOH$ $\rightarrow CH_3COO^-Na^+ + H_2O$	Mix aqueous solutions
Sodium hydrogencarbonate	$CH_3COOH + NaHCO_3$ $\rightarrow CH_3COONa + CO_2 + H_2O$	Add to a solution of sodium hydrogencarbonate and detect gaseous CO_2
Alcohols (e.g. ethanol)	$CH_3COOH + C_2H_5OH$ $\rightleftharpoons CH_3COOC_2H_5 + H_2O$	Warm with a few drops of concentrated sulfuric acid
Phosphorus pentachloride	$CH_3COOH + PCl_5$ $\rightarrow CH_3COCl + HCl + POCl_3$	Mix dry reagents at room temperature
Lithium aluminium hydride (reduction)	$CH_3COOH + 4[H]$ $\rightarrow CH_3CH_2OH + H_2O$	Dry ether solution, then add dilute hydrochloric acid

Acid chlorides

In the table below, ethanoyl chloride is used as an example.

Reagent	Equation	Conditions
Water	$CH_3COCl + H_2O \rightarrow CH_3COOH + HCl$	Mix
Alcohols (e.g. ethanol)	$CH_3COCl + C_2H_5OH \rightarrow CH_3COOC_2H_5 + HCl$	Dry reagents
Ammonia	$CH_3COCl + 2NH_3 \rightarrow CH_3CONH_2 + NH_4Cl$	Concentrated ammonia
Amines (e.g. ethylamine)	$CH_3COCl + C_2H_5NH_2 \rightarrow CH_3CONHC_2H_5 + HCl$	Mix

Esters

In the table below, ethyl ethanoate is used as the example.

Reagent	Equation	Conditions
Sodium hydroxide	$CH_3COOC_2H_5 + NaOH$ $\rightarrow CH_3COO^-Na^+ + C_2H_5OH$	Heat under reflux in aqueous solution
Inorganic acids	$CH_3COOC_2H_5 + H_2O$ $\rightleftharpoons CH_3COOH + C_2H_5OH$	Heat under reflux with dilute sulfuric acid catalyst
Carboxylic acids (transesterification)	$CH_3COOC_2H_5 + HCOOH$ $\rightleftharpoons HCOOC_2H_5 + CH_3COOH$	Acid catalyst
Alcohols (transesterification)	$CH_3COOC_2H_5 + CH_3OH$ $\rightleftharpoons CH_3COOCH_3 + C_2H_5OH$	Acid catalyst

Amines

In the table below, ethylamine is used as the example.

Reagent	Equation	Conditions
Acids (e.g. hydrochloric)	$C_2H_5NH_2 + HCl \rightarrow C_2H_5NH_3{}^+Cl^-$	Mix aqueous solutions
Acid chlorides (e.g. ethanoyl chloride)	$C_2H_5NH_2 + CH_3COCl$ $\rightarrow CH_3CONHC_2H_5 + HCl$	Mix
Halogenoalkanes	$C_2H_5NH_2 + C_2H_5I \rightarrow (C_2H_5)_2NH + HI$	Warm in ethanolic solution
Copper(II) ions	$4C_2H_5NH_2 + [Cu(H_2O)_6]^{2+}$ $\rightarrow [Cu(C_2H_5NH_2)_4(H_2O)_2]^{2+} + 4H_2O$	Mix aqueous solutions

Nitriles

In the table below, ethanenitrile is used as an example.

Reagent	Equation	Conditions
Dilute hydrochloric acid (hydrolysis)	$CH_3CN + 2H_2O + HCl$ $\rightarrow CH_3COOH + NH_4Cl$	Heat under reflux
Lithium aluminium hydride (reduction)	$CH_3CN + 4[H] \rightarrow C_2H_5NH_2$	Dry ether solution, then add dilute hydrochloric acid

Aromatic chemistry

Benzene

Reagent	Equation	Conditions
Nitric acid	$C_6H_6 + HNO_3 \rightarrow C_6H_5NO_2 + H_2O$	Mix concentrated nitric and sulfuric acids at 50°C
Bromine	$C_6H_6 + Br_2 \rightarrow C_6H_5Br + HBr$	Liquid bromine and iron
Fuming sulfuric acid	$C_6H_6 + SO_3 \rightarrow C_6H_5SO_3H$	Warm
Acid chlorides, e.g. ethanoyl chloride	$C_6H_6 + CH_3COCl$ $\rightarrow C_6H_5COCH_3 + HCl$	Anhydrous aluminium chloride catalyst
Halogenoalkanes	$C_6H_6 + C_2H_5Br \rightarrow C_6H_5C_2H_5 + HBr$	Anhydrous aluminium chloride catalyst

Phenol

OH

$+\ 3Br_2 \longrightarrow$

Bromine
water

OH

Br ⟶ Br

Br

$+\ 3HBr$

Mix and observe white
precipitate

Dilute nitric acid at room temperature

OH

$+\quad HNO_3(aq) \longrightarrow$

OH

NO_2

and

OH

NO_2

NO_2

$+\quad H_2O$

Phenylamine

Reagent	Equation	Conditions
Acid	$C_6H_5NH_2 + H^+ \rightarrow C_6H_5NH_3^+$	Add dilute acid
Acid chloride	$C_6H_5NH_2 + CH_3COCl$ $\rightarrow CH_3CONHC_6H_5 + HCl$	Mix
With nitrous acid at 5°C	$C_6H_5NH_2 + HNO_2 + H^+$ $\rightarrow C_6H_5N_2^+ + 2H_2O$	Nitrous acid made in situ by adding dilute hydrochloric acid to aqueous sodium nitrite

Questions
&
Answers

This section contains multiple-choice and structured questions similar to those you can expect to find in Unit Test 5. The questions given here are not balanced in terms of types of questions or level of demand.

Examiner comments

The sample answers are followed by examiner comments, preceded by the icon . These comments may explain the correct answer or point out common errors.

Multiple-choice questions

Each question or incomplete statement is followed by four suggested answers, A, B, C or D. Select the *best* answer in each case. The answers are given, with some commentary, after question 15.

1 When the standard electrode potential of zinc is measured, a reference electrode is needed because:

   A the potential between a metal and its solution cannot be measured directly

   B a standard hydrogen electrode has a potential of zero

   C zinc is the cathode

   D a pressure of 1 atm must be used

2 If the standard electrode potential of a cell, $E^{\ominus}_{cell}$, is +0.10 V, the reaction:

   A is not thermodynamically feasible, as the value of $E^{\ominus}_{cell}$ is too small

   B will always take place

   C will take place under standard conditions, if the activation energy is not too high

   D will have a negative value of ΔS_{total}

3 Which of the following can act as a bidentate ligand?

   A NH_2NH_2 **B** $NH_2(CH_2)_2NH_2$

   C $ClCH_2Cl$ **D** $Cl(CH_2)_2Cl$

4 Which of the following is the electronic configuration of a stable, coloured transition-metal ion?

   A [Ar] $3d^{10}\ 4s^1$ **B** [Ar] $3d^{10}\ 4s^0$

   C [Ar] $3d^6\ 4s^1$ **D** [Ar] $3d^6\ 4s^0$

5 When solid hydrated copper sulfate ($CuSO_4.5H_2O$) is heated, it produces steam and white anhydrous copper sulfate ($CuSO_4$). This compound is white because:

   A copper has an electron configuration of [Ar] $3d^{10}\ 4s^1$

   B the copper ion in $CuSO_4$ has a full *d*-shell

   C the copper ion no longer has any ligands and so the *d*-orbitals are not split

   D the copper ions are reduced to Cu^+ ions

6 An isomer of C_5H_8O gave a yellow precipitate with iodine and alkali, decolorised bromine water and gave a yellow precipitate with 2,4-dinitrophenylhydrazine. The isomer is:

A $CH_2=C=CHCH(OH)CH_3$

B $CH_3CH=CHCH_2CHO$

C $CH_2=CHCH_2COCH_3$

D $CH_2=CHCOCH_2CH_3$

7 The electrophile in the Friedel–Crafts reaction between benzene and ethanoyl chloride is:

A $CH_3\overset{+}{C}O$

B $CH_3\overset{+}{C}O^-$

C $CH_3\overset{+}{C}OCl^-$

D $CH_3\overset{\delta+}{C}\overset{\delta-}{O}$

8 When phenol reacts with bromine water, the product(s) are:

A

B

+ 3HBr

C

D

+ HBr

9 Which of the following is **not** an amide?

A CH_3CONH_2

B $CH_3CONHC_2H_5$

C $C_6H_5CONH_2$

D $CH_3COCH_2NH_2$

10 Which statement about amines, for example ethylamine, is **not** true?

A They react with bases.

B They are soluble in water.

C They react with acid chlorides.

D They form complex cations with hydrated copper(II) ions.

11 What are the reagents and conditions for the following two-step synthesis?

step 1   step 2
nitrobenzene → phenylamine → benzenediazonium chloride

	Step I	Step 2
A	Tin and concentrated hydrochloric acid	Sodium nitrite and hydrochloric acid below 0°C
B	Iron and concentrated hydrochloric acid	Sodium nitrite and hydrochloric acid between 0°C and 10°C
C	Concentrated hydrochloric acid	Sodium nitrite and hydrochloric acid above 10°C
D	Hot acidified potassium dichromate(VI)	Sodium nitrite and hydrochloric acid heated under reflux

12 A polyamide such as Kevlar®:

A is water soluble because of hydrogen bonds between the —NH groups and water

B is made from an amine such as ethylamine and a dicarboxylic acid such as benzene-1,3-dicarboxylic acid

C is a solid with hydrogen bonding between strands

D is biodegradable

13 Poly(ethenol) is:

A made by polymerising ethenol at room temperature

B made by polymerising ethenol using a Ziegler–Natta catalyst

C made by a hydrolysis of polyvinyl acetate with dilute acid

D insoluble in water because it cannot form enough hydrogen bonds with water molecules

14 Which of the following compounds has an unsplit peak (a singlet) in its NMR spectrum?

A CH_3CHO B $(CH_3)_3CH$

C $(CH_3)_2CHCHO$ D $CH_3COCH_2CH_3$

15 Heating under reflux is used if the reaction is slow at room temperature and:

A the product has a low boiling temperature

B one of the reactants has a low boiling temperature

C the reactants and products have similar boiling temperatures

D the reactants and product form separate layers

Answers

1 A

📝 Zinc ions go into solution and the electrons remain on the zinc rod, thus setting up a potential difference. However, this cannot be measured directly as any connection to a voltmeter using a metal wire dipping into the solution would alter the potential. A salt bridge has to be used which is then connected to a reference electrode. Options **B** and **D** are true statements but neither is the reason why a reference electrode has to be used.

Option **C** is incorrect, as the zinc becomes oxidised and is, therefore, the anode.

2 C

📝 It is a common misunderstanding that the cell potential has to be above a value, such as $> +0.03\,V$ or $> +0.3\,V$. As long as the cell potential is positive, the reaction is thermodynamically feasible. Therefore, option **A** is incorrect. Option **B** is incorrect because the reaction may not happen for kinetic reasons. Option **D** is incorrect because a positive value for the cell potential means that ΔS_{total} is also positive.

3 B

📝 Hydrazine (NH_2NH_2) has a lone pair of electrons on each nitrogen atom. If it formed a bidentate ligand, this would involve a triangle of the two nitrogen atoms and the metal ion. Such a triangular shape would cause too much strain in the bonds. Therefore, option **A** is incorrect. Option **B** is correct because 1,2-diaminoethane has two lone pairs of electrons, one on each nitrogen atom, and it forms a five-membered ring with the metal ion and this is stable. A covalently bound chlorine atom is not a ligand (although chlorine ions can be), so options **C** and **D** are incorrect.

4 D

📝 This is the electronic configuration of a Fe^{2+} ion. Option **A** is the configuration of a copper atom — note that the question asks about a transition metal *ion*. Option **B** is the configuration of Zn^{2+} or Cu^+. Zinc is not a transition metal, as its ion does not have an unpaired *d*-electron and copper(I) ions are not coloured. Option **C** is the configuration for a Fe^+ ion, which does not exist.

5 C

📝 Hydrated copper sulfate has four water ligands. When it is heated, these are lost. As there are now no ligands, there is no splitting of the *d*-orbitals. Option **A** is a true statement about a copper atom, but is irrelevant to this question. Option **B** is incorrect as the copper ion in $CuSO_4$ is $2+$ and has the electron configuration $[Ar]\,3d^9$. Option **D** is incorrect because hydrated copper(II) sulfate is not reduced on heating.

6 C

📝 The iodoform reaction tests for a $CH_3C=O$ or a $CH_3CH(OH)$ group. Bromine water tests for a $C=C$ group and Brady's reagent for a $C=O$ group. The question is best attempted by putting a ✓ or ✗ for each reaction.

	CHI₃ test	Br₂(aq)	2,4-DNP
A	✓	✓	✗
B	✗	✓	✓
C	✓	✓	✓
D	✗	✓	✓

7 A

📝 The Cl is removed along with the σ-bond electrons, leaving the positive charge on the carbonyl carbon atom. Options **B**, **C** and **D** are simply incorrect.

8 B

📝 Phenol and bromine react in a substitution reaction, so there must be two products. Thus options **A** and **C** are incorrect. The oxygen atom activates the benzene ring and three bromine atoms substitute into the ring in the 2, 4 and 6 positions. Thus **D** is incorrect.

9 D

📝 An amide has a nitrogen atom attached to the carbon of a $C=O$ group. Option **A** has a $C=ONH_2$ group and is ethanamide. Option **B** is a substituted amide having a $C=ONH$ group and is *N*-ethylethanamide. Option **C** is also an amide — benzamide. In option **D**, the $C=O$ and NH_2 groups are separated by a CH_2 group. Therefore, this compound is a ketone and an amine; it is not an amide. Thus **D** is the correct answer to this negative question.

10 A

📝 Amines are bases and not acids. Therefore, they do not react with other bases, so option **A** is false and is the correct answer to this negative question. Amines are soluble in water, as they can form hydrogen bonds with H_2O molecules. They react with acid chlorides to form substituted amides. An amine molecule has a lone pair of electrons on the nitrogen atom that can act as a ligand with transition metal ions such as copper(II) ions. Therefore, options **B**, **C** and **D** are all true statements.

11 B

The correct reagents for the reduction of nitrobenzene are either tin or iron plus concentrated hydrochloric acid. So, for step 1, options **A** and **B** are correct and options **C** (no reducing agent) and **D** (dichromate(VI) ions are an oxidising agent) are incorrect. For step 2, the temperature must not be above 10°C (another reason why option **C** is wrong) because this would cause the diazonium ions to decompose; it must be above 0°C or the reaction will be too slow. Therefore, option **A** is incorrect. Only option **B** has the correct reagents and conditions.

12 C

Kevlar is a polyamide and so forms strong hydrogen bonds between strands with the δ+ hydrogen of the –NH forming a hydrogen bond with the lone pair of the δ– oxygen atom in a C=O group in another strand. In addition, the benzene rings form van der Waals attractions between strands, making Kevlar stronger than other polyamides. Option **A** is incorrect because the benzene rings are hydrophobic and the interstrand hydrogen bonding is so strong that Kevlar is not water-soluble. The monomers of a polyamide must have two functional groups. Ethylamine has only one functional group, so it cannot form polymers. Therefore, option **B** is incorrect. Kevlar is not a natural polymer and so is not biodegradable. Therefore, option **D** is also incorrect.

13 C

Ethenol (CH_2=CHOH) does not exist. Attempts to make it produce ethanal (CH_3CHO), so options **A** and **B** are incorrect. Poly(ethenol) is water-soluble, so option **D** is wrong. It is manufactured either by transesterification with methanol or by the hydrolysis of polyvinyl acetate (($H_2CCH(OOCCH_3$))$_n$) so **C** is the correct answer.

14 D

A singlet can be due to the hydrogen of an OH group, or to the hydrogen atoms on a carbon atom that does not have an adjacent carbon atom with a hydrogen atom attached. Thus all alcohols and carboxylic acids have a singlet, as do methyl ketones, which have a CH_3CO group and thus the methyl hydrogen atoms have no neighbouring hydrogen atoms to cause splitting. The CH_3 hydrogen atoms in butanone (**D**) (on the left as written in the question) have a C=O group as their neighbour and so this peak will not be split. Ethanal (**A**) has two peaks, one split into four and the other into two. Methylpropane (**B**) has two peaks, one split into ten and the other into two. Methylpropanal (**C**) has three peaks, one split into seven and two split into two.

15 B

Be careful here: it is the reactant not the product that must be prevented from boiling off, so option **A** is incorrect The responses in **C** and **D** have nothing whatsoever to do with heating under reflux.

Structured questions

Responses to questions with longer answers may have an examiner's comment preceded by the icon 🄴. Some of the comments highlight common errors made by candidates who produce work of C-grade standard or lower.

Question parts marked with an asterisk (*) test 'quality of written communication' (QWC).

It is worth reading through the whole question before attempting to answer it.

Question 16

[handwritten annotations: carboxylic acid ↓ PCl₅ → acyl chloride → ammonia]

(a) **Ethanamide (CH_3CONH_2) can be prepared in the laboratory from ethanol. Outline how this could be carried out, stating the reagents and conditions needed and identifying all intermediate compounds formed.** (6 marks)

(b) **Tylenol is a secondary amide and a phenol. It is used as a painkiller in the USA. Predict the organic product of its reaction with:**
  (i)   **bromine water** (1 mark)
  (ii) **aqueous sodium hydroxide at room temperature** *[handwritten: O⁻Na⁺]* (1 mark)

[handwritten structure: CH₃ ... O=C ... C–C–N–H with H₀]

[structure of Tylenol: O=C with CH₃ branch, NH connecting to benzene ring with OH]

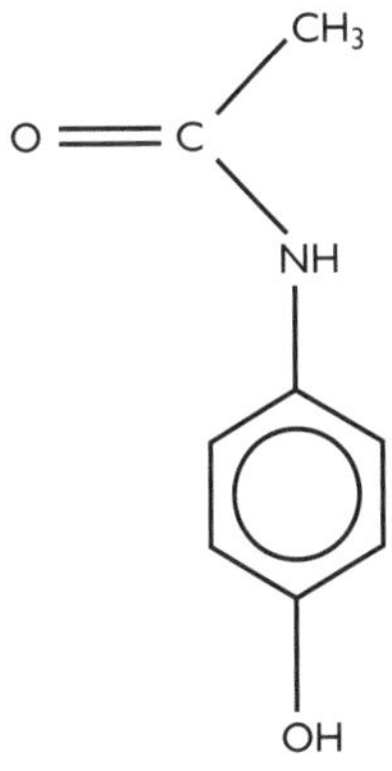

(c) **An organic compound Z contains only carbon, hydrogen, nitrogen and oxygen. When 1.00 g of Z was burnt, 1.48 g of carbon dioxide, 0.71 g of water and 0.52 g of nitrogen dioxide (NO_2) were produced.**
  (i)   **Calculate the percentage by mass of the elements in Z.** (2 marks)
  (ii) **Use your answer to (i) to show that Z has the empirical formula $C_3H_7O_2N$.** *[handwritten: H₇ ?]* (2 marks)

(d) **The molecular formula of compound Z is the same as its empirical formula. Compound Z melts at a high temperature, is soluble in water and reacts with both acids and alkalis. When heated with ninhydrin, a dark colour is produced.**
  (i)   **Draw the structural formula of Z. Z has a chiral centre.** (2 marks)
  *(ii) **Explain why compound Z has a high melting temperature and why it is soluble in water.** (4 marks)
  (iii) **Write ionic equations to show the reactions of compound Z with H^+(aq) ions and with OH^-(aq) ions.** (2 marks)

Total: 20 marks

Answers to question 16

(a) Step 1: $C_2H_5OH \rightarrow CH_3COOH$ ✓ (ethanoic acid)
Reagents: acidified potassium dichromate ✓ ($H^+/Cr_2O_7^{2-}$)
Conditions: heat under reflux ✓
Step 2: $CH_3COOH \rightarrow CH_3COCl$ ✓ (ethanoyl chloride)
Reagent: phosphorus pentachloride ✓ (PCl_5)
Step 3: $CH_3COCl \rightarrow CH_3CONH_2$
Reagent: ammonia solution ✓ (NH_3)

✎ Alternative reagents for step 2 are PCl_3 or $SOCl_2$.

(b) (i) ✓

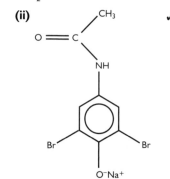

(ii) ✓

✎ Tylenol is a phenol. It reacts with bromine water to form a white precipitate.

✎ As Tylenol is a phenol, it reacts as an acid with aqueous sodium hydroxide. The amide group is not hydrolysed at room temperature.

(c) (i) mass of carbon = $1.48 \times \dfrac{12.0}{44.0} = 0.404\,g$ % carbon = 40.4%

mass of hydrogen = $0.71 \times \dfrac{2.0}{18.0} = 0.789\,g$ % hydrogen = 7.89%

mass of nitrogen = $0.52 \times \dfrac{14.0}{46.0} = 0.158\,g$ % nitrogen = 15.8%

% oxygen = $100 - 40.4 - 7.89 - 15.8 = 35.91\%$

✎ Note that to obtain the mass of hydrogen, the mass of water has to be multiplied by 2, because there are two hydrogen atoms in each water molecule. In this example the percentage is 100 times the mass because the initial mass of Z was 1.00 g.

(ii)

Element	%	Divide by Ar	Divide by smallest
Carbon	40.4	$\dfrac{40.4}{12.0} = 3.37$	$\dfrac{3.37}{1.13} = 3.0$
Hydrogen	7.89	$\dfrac{7.89}{1.0} = 7.89$	$\dfrac{7.89}{1.13} = 7.0$
Nitrogen	15.8	$\dfrac{15.8}{14.0} = 1.13$	$\dfrac{1.13}{1.13} = 1.0$
Oxygen	35.91	$\dfrac{35.91}{16.0} = 2.24$	$\dfrac{2.24}{1.13} = 2.0$

This is consistent with an empirical formula of $C_3H_7O_2N$. ✓

(d) (i) The formula is:

$$CH_3$$

$$H_2N\text{---}\overset{*}{C}\text{---}COOH \quad or \quad H_3\overset{+}{N}\text{---}\overset{*}{C}\text{---}COO^-$$

$$H \qquad\qquad\qquad H$$

There is 1 mark for either an H_2N/H_3N^+ group or a $COOH$/COO^- group and 1 mark for the remainder. The chiral centre is marked with an asterisk.

Ninhydrin reacts with amino acids to form a dark coloration. This is how the positions of different amino acids are identified in thin-layer chromatography.

(ii) Compound Z has a high melting temperature because it forms a zwitterion, $^+H_3NCH(CH_3)COO^-$ ✓. There are strong ionic forces of attraction between different zwitterions. This means that a lot of energy is required to separate zwitterions and therefore the melting temperature is high ✓.

It is soluble in water because hydrogen bonds form between the COO^- groups and the δ+ hydrogen atoms in water ✓ and between the ^+H_3N groups and the lone pairs of electrons on the δ− oxygen atoms in water ✓.

(iii) Either $\quad H_2NCH(CH_3)COOH + H^+ \rightarrow {}^+H_3NCH(CH_3)COOH$

or $\qquad {}^+H_3NCH(CH_3)COO^- + H^+ \rightarrow {}^+H_3NCH(CH_3)COOH$ ✓

and $\quad$ either $H_2NCH(CH_3)COOH + OH^- \rightarrow H_2NCH(CH_3)COO^- + H_2O$

or $\qquad {}^+H_3NCH(CH_3)COO^- + OH^- \rightarrow H_2NCH(CH_3)COO^- + H_2O$ ✓

Question 17

(a) $Cr^{3+}(aq) + e^- \rightleftharpoons Cr^{2+}(aq)$ $\qquad\qquad\qquad\qquad E^\ominus = -0.41\,V$
$Cr_2O_7{}^{2-}(aq) + 14H^+(aq) + 6e^- \rightleftharpoons 2Cr^{3+}(aq) + 7H_2O(l)$ $\quad E^\ominus = +1.33\,V$
$S(s) + 2H^+(aq) + 2e^- \rightleftharpoons H_2S(aq)$ $\qquad\qquad\qquad E^\ominus = +0.14\,V$

(i) Hydrogen sulfide (H_2S) reduces dichromate(VI) ions in acid solution. Is the product Cr^{3+} ions or Cr^{2+} ions? Justify your answer, calculating the $E^\ominus_{cell}$ values for possible reductions. $\qquad$ (4 marks)

(ii) Write the ionic equation for the reaction that takes place between hydrogen sulfide and dichromate(VI) ions in acid solution. $\qquad$ (2 marks)

(iii) Define disproportionation. $\qquad$ (1 mark)

*(iv) Explain in terms of $E^\ominus$ values, whether $Cr^{3+}(aq)$ ions will disproportionate. $\qquad$ (2 marks)

(b) The water from volcanic hot springs often contains significant amounts of the poisonous gas hydrogen sulfide. The amount present can be measured by titration against a standard solution of acidified potassium manganate(VII).

$$5H_2S + 2MnO_4{}^- + 6H^+ \rightarrow 2Mn^{2+} + 5S + 8H_2O$$

A portion of hot spring water of volume 250 cm³ was taken and 50.0 cm³ portions were titrated against acidified 0.0100 mol dm⁻³ potassium manganate(VII) solution. The mean titre was 7.55 cm³. Calculate the volume of hydrogen sulfide dissolved in the 250 cm³ sample of hot spring water. (1 mol of gas, under the conditions of the experiment, has a volume of 24 000 cm³) $\qquad$ (4 marks)

$\qquad\qquad\qquad\qquad\qquad\qquad\qquad\qquad\qquad\qquad$ **Total: 13 marks**

Answers to question 17

(a) (i)   Reduction to Cr^{3+}:

The third equation has to be reversed and multiplied by 3:

$Cr_2O_7^{2-}(aq) + 14H^+(aq) + 6e^- \rightleftharpoons 2Cr^{3+}(aq) + 7H_2O(l)$ $E^{\ominus} = +1.33\,V$

$3H_2S(aq) \rightleftharpoons 6H^+(aq) + 3S(s) + 6e$ $E^{\ominus} = -(10.14\,V) = -0.14\,V$

The value of $E^{\ominus}_{cell}$ is $(+1.33) + (-0.14) = +1.19\,V$ ✓

This is positive, so hydrogen sulfide will reduce chromium(VI) ions to chromium(III) ions ✓.

📝 Remember that when a redox half-equation is reversed, the sign of $E^{\ominus}$ must be altered; when it is multiplied by a number, the $E^{\ominus}$ value is unaltered.

The equations need not be written here, but it is safer to do so. The alternative is to realise that the $Cr_2O_7^{2-}/Cr^{3+}$ equation is the right way round but the S/H_2S equation has to be reversed. Therefore the value of:

$E^{\ominus}_{cell}$ = (the value of $Cr_2O_7^{2-}/Cr^{3+}$ half-equation) − (the value of S/H_2S half-equation)

$= +1.33 - (+0.14) = +1.19\,V$

Reduction to Cr^{2+}:

The first equation has to be multiplied by 2 and the third equation reversed:

$2Cr^{3+}(aq) + 2e^- \rightleftharpoons 2Cr^{2+}(aq)$ $E^{\ominus} = -0.41\,V$

$H_2S(aq) \rightleftharpoons 2H^+(aq) + S(s) + 2e^-$ $E^{\ominus} = -(+0.14\,V) = -0.14\,V$

The value of $E^{\ominus}_{cell}$ is $-0.41 + (-0.14) = -0.55\,V$ ✓

This is negative so hydrogen sulfide will reduce chromium(VI) ions to chromium(III) ions but no further ✓.

(ii)   The overall equation is obtained by adding:

$Cr_2O_7^{2-}(aq) + 14H^+(aq) + 6e^- \rightarrow 2Cr^{3+}(aq) + 7H_2O(l)$

$3H_2S(aq) \rightarrow 6H^+(aq) + 3S(s) + 6e^-$ ✓

$Cr_2O_7^{2-}(aq) + 8H^+(aq) + 3H_2S \rightarrow 2Cr^{3+}(aq) + 7H_2O(l) + 3S(s)$ ✓

📝 You must remember to cancel the H^+ ions: $14H^+$ on the left-hand side and $6H^+$ on the right-hand side cancel to leave $8H^+$ on the left.

(iii)   Disproportionation is when an element in a single species is simultaneously oxidised and reduced. ✓

(iv)   The disproportionation reaction would involve Cr^{3+} ions being reduced to Cr^{2+} ions (first equation in the data) and Cr^{3+} ions being oxidised to $Cr_2O_7^{2-}$ ions (the second equation reversed). ✓

The $E^{\ominus}_{cell}$ for this is ($E^{\ominus}$ for the first half-equation) − ($E^{\ominus}$ for the second half-equation):

$E^{\ominus}_{cell} = (-0.41) - (+1.33) = -1.74\,V$. This value is negative and so disproportionation of Cr^{3+} ions does not occur. ✓

(b) amount (moles) of MnO_4^- = concentration × volume

$$= 0.0100 \, mol \, dm^{-3} \times 0.00755 \, dm^3$$

$$= 7.55 \times 10^{-5} \, mol \checkmark$$

amount (moles) of H_2S in $50 \, cm^3 = \frac{5}{2} \times 7.55 \times 10^{-5} = 1.8875 \times 10^{-4} \, mol \checkmark$

amount (moles) in $\underline{250 \, cm^3} = \frac{250}{50} \times 1.8875 \times 10^{-4} = 9.4375 \times 10^{-4} \, mol \checkmark$

volume of hydrogen sulfide dissolved $= 9.4375 \times 10^{-4} \times 24\,000 = 22.7 \, cm^3 \checkmark$

Question 18

(a) Explain, with the aid of an equation, why hydrated chromium(III) ions in solution are acidic. (2 marks)

(b) State what you would see when aqueous sodium hydroxide is added steadily until in excess to a solution of hydrated chromium(III) ions. Give the equations for the reactions. (4 marks)

***(c) EDTA reacts with hydrated chromium(III) ions to form a polydentate complex ion in an exothermic reaction:**

$EDTA^{4-}(aq) + [Cr(H_2O)_6]^{3+}(aq) \rightarrow [Cr(EDTA)]^-(aq) + 6H_2O(l)$

Explain why ΔS_{system} is positive for this reaction. (2 marks)

***(d) The EDTA complex ion is coloured. Explain why it is coloured.** (3 marks)

Total: 11 marks

Answers to question 18

(a) The hydrated ions are reversibly deprotonated by water forming H_3O^+ ions which make the solution acidic. ✓

$$[Cr(H_2O)_6]^{3+}(aq) + H_2O(l) \rightleftharpoons [Cr(H_2O)_5OH]^{2+}(aq) + H_3O^+(aq) \checkmark$$

(b) At first a dirty green precipitate ✓ is formed.

$$[Cr(H_2O)_6]^{3+}(aq) + 3OH^-(aq) \rightarrow [Cr(H_2O)_3(OH)_3](s) + 3H_2O(l) \checkmark$$

This then dissolves in excess sodium hydroxide to form a green solution. ✓

$$[Cr(H_2O)_3(OH)_3](s) + 3OH^-(aq) \rightarrow [Cr(OH)_6]^{3-} + 3H_2O(l) \checkmark$$

(c) The reaction with EDTA involves two particles reacting to form seven particles, all in the aqueous phase. ✓ This is a significant increase in disorder and so ΔS_{system} is positive. ✓

(d) The six pairs of electrons forming the bonds between EDTA and the Cr^{3+} ion cause the d-orbitals in the chromium ion to split into two levels. ✓ When white light is shone onto the complex, some of the frequencies are absorbed ✓ and an electron is promoted from the lower level of the split d-orbitals to a higher level ✓. This causes the ion to have the complementary colour to that of the wavelength absorbed.

Question 19

Arenes, such as benzene and its derivatives nitrobenzene and phenol, react with electrophiles.

(a) Nitrobenzene can be prepared from benzene. Give the reagents and conditions and write the equation for this reaction. Why must the temperature be controlled? (5 marks)

(b) Benzene also reacts with ethanoyl chloride in a Friedel–Crafts reaction. Give the mechanism for this reaction, including the production of the electrophile. (4 marks)

(c) Phenol reacts with bromine water.
  (i)   Write the equation for this reaction. (1 mark)
  (ii) Explain why this reaction does not need a catalyst and is much faster than the reaction of bromine with benzene. (3 marks)

Total: 13 marks

Answers to question 19

(a) The reagents are concentrated nitric acid ✓ and concentrated sulfuric acid ✓. The conditions are heat to 50°C ✓ (At a higher temperature some dinitrobenzene is produced and at a lower temperature the reaction is too slow.) The equation is:

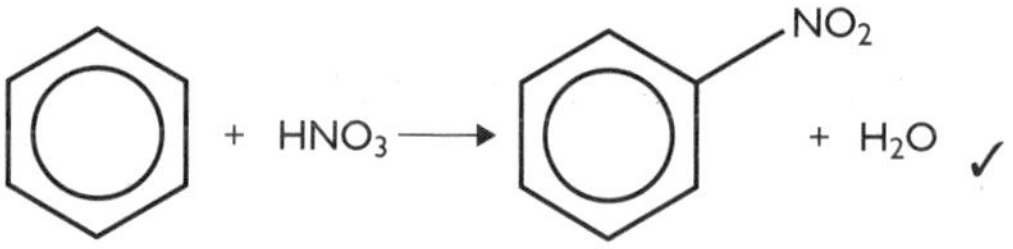

$+ HNO_3 \longrightarrow$ ... $+ H_2O$ ✓

(b) Formation of electrophile

$$CH_3COCl + AlCl_3 \rightarrow CH_3C^+=O + AlCl_4^-$$

Step I

$CH_3C^+=O \longrightarrow$

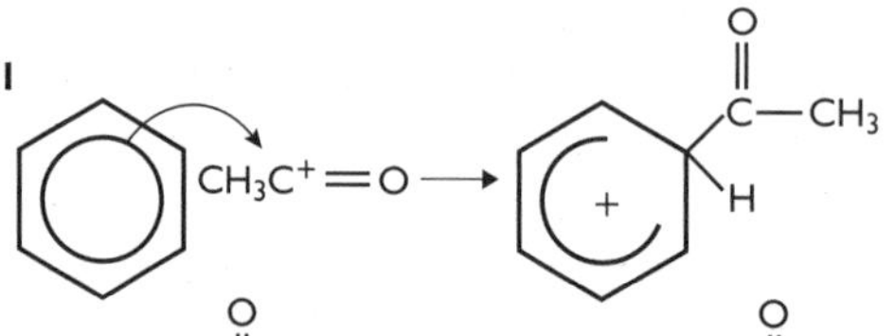

Step 2

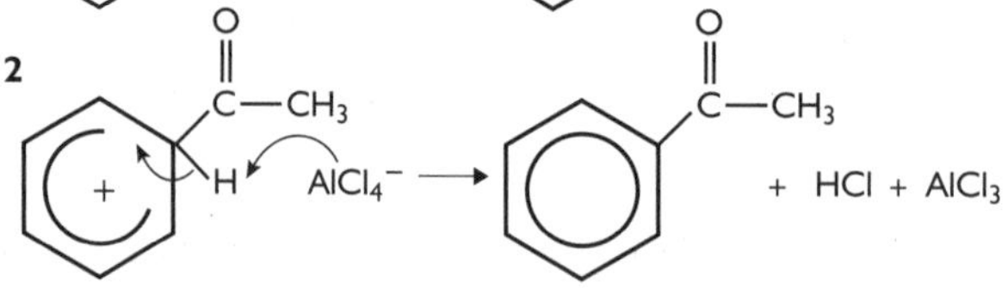

$AlCl_4^- \longrightarrow$ ... $+ HCl + AlCl_3$

(c) (i)

$+ 3Br_2 \longrightarrow$ ... $+ 3HBr$

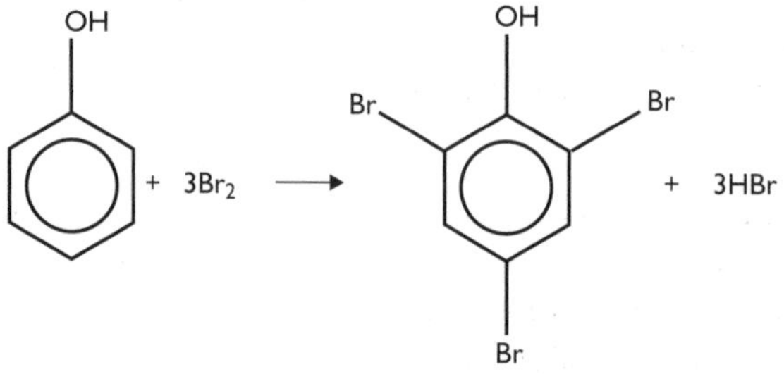

(ii)   The lone pair of electrons (the $2p_z$ electrons) on the oxygen atom in phenol becomes partly delocalised into the benzene ring π-bond. ✓ This increases the electron density in the ring and makes it more susceptible to electrophilic attack ✓ and so has a lower activation energy, making the reaction faster. ✓

Contemporary context question

Question 20

Read the passage carefully and answer the questions that follow.

Electromagnetic radiation

Electromagnetic radiation is important in chemistry in a number of ways.

Radio waves of frequency around 100–400 MHz are used in NMR spectroscopy. The substance under investigation is placed in a very strong magnetic field. This causes the energy levels of spinning hydrogen nuclei to split into two. The peaks in the NMR spectrum measure the difference between these two levels, which varies according to the exact environment of the hydrogen nuclei. The peaks are also split by the hydrogen atoms on neighbouring carbon atoms; the splitting follows the $(n + 1)$ rule. Microwaves have a higher frequency. This radiation is only absorbed by polar molecules. The molecules then rotate faster gaining rotational energy. On collision with another molecule, this energy is converted into kinetic energy, in the form of heat. Use of this is made in the pharmaceutical industry where microwaves are used to heat up reaction mixtures.

Infrared spectroscopy uses radiation of a higher frequency still. It detects the presence of functional groups and can be used to identify unknown compounds. When infrared radiation is absorbed by a molecule, one or more of its bonds bend or stretch. Absorption only takes place if the stretching or bending causes a change in dipole moment.

UV light is the highest frequency radiation normally used in chemistry. It causes bonds in a molecule to break homolytically, forming free radicals. When a mixture of chlorine and methane is exposed to a short pulse of UV radiation, chlorine radicals are produced. These then take part in a chain reaction, the first step being the removal of a hydrogen atom from methane to form hydrogen chloride and a methyl radical.

(a) Place the four types of electromagnetic radiation discussed in the passage in order of decreasing frequency.   (1 mark)

*(b) Explain why the air inside a microwave oven does not heat up when a cup of water is being heated in the oven.   (1 mark)

*(c) (i)   State and explain the difference between the high-resolution NMR spectra of ethanol (CH_3CH_2OH) and ethanamide (CH_3CONH_2).   (5 marks)

   (ii) The infrared spectra of ethanol and ethanamide compounds are shown below as spectrum X and spectrum Y.

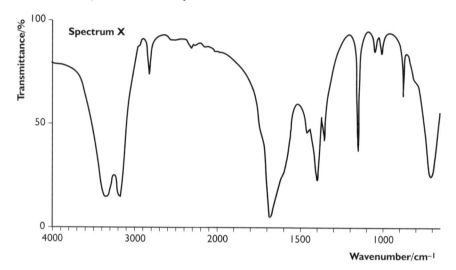

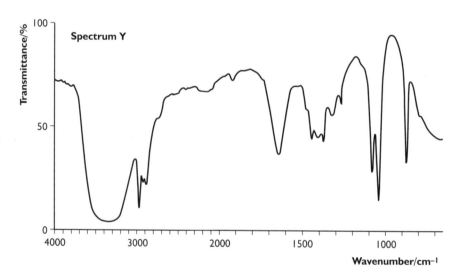

Identify the spectrum of ethanamide, giving reasons. (You should consult the Edexcel data booklet.)   (3 marks)

(d) When chlorine gas is bubbled into liquid methylbenzene ($C_6H_5CH_3$) in the presence of UV radiation, a free radical substitution reaction takes place. One of the products is chloromethylbenzene ($C_6H_5CH_2Cl$). The mechanism of this reaction is similar to that between methane and chlorine.

(i) Write the mechanism of the reaction between chlorine gas and methylbenzene, showing the initiation step and two propagation steps. (3 marks)

(ii) Some 1,2-diphenylethane ($C_6H_5CH_2CH_2C_6H_5$) is formed in this reaction. Explain how this is evidence for the mechanism that you have written in (i). (2 marks)

*(iii) Suggest how a pure dry sample of chloromethylbenzene could be extracted from the reaction mixture. (5 marks)

	Density/ g cm^{-3}	Boiling temperature/°C	Solubility in water
Chloromethylbenzene	1.1	179	Insoluble
Methylbenzene	0.87	111	Insoluble

Total: 20 marks

Answers to question 20

(a) UV > IR > microwaves > radio waves ✓

(b) The air consists mostly of oxygen and nitrogen. These are non-polar molecules and so do not absorb microwaves ✓.

(c) (i) Ethanol has the formula CH_3CH_2OH and has hydrogen nuclei in three chemical environments. Its NMR spectrum has three different peaks in the ratio 3:2:1. ✓ The peak due to the hydrogen atoms in the CH_3 group is split into three and those due to the hydrogen atoms in the CH_2 group are split into four. ✓ The peak due to the hydrogen in the OH group is not split. ✓ Ethanamide has the formula CH_3CONH_2 and has hydrogen nuclei in two chemical environments. Its spectrum has two peaks in the ratio 3:2 ✓ and neither of the peaks is split ✓.

📝 Do not forget to describe for both molecules the number of peaks, their relative heights and the extent of splitting due to spin coupling ($n + 1$) rule. The peaks in the ethanamide spectrum are not split, because there is no hydrogen atom on the carbon atoms adjacent to the CH_3 and NH_2 groups.

(ii) Both spectra have a broad peak at around $3350\,cm^{-1}$. These are caused by the O—H bond in ethanol and the N—H bond in ethanamide. ✓ Spectrum X has a strong peak at $1680\,cm^{-1}$. This is typical of the C=O bond in amides ($1700–1630\,cm^{-1}$) ✓, so spectrum X is that of ethanamide and spectrum Y is that of ethanol ✓.

(d) (i) Initiation step:

$Cl_2 \xrightarrow{UV} 2Cl\bullet$ ✓

Propagation steps:

$Cl\bullet + C_6H_5CH_3 \rightarrow C_6H_5CH_2\bullet + HCl$ ✓

$C_6H_5CH_2\bullet + Cl_2 \rightarrow C_6H_5CH_2Cl + Cl\bullet$ ✓

(ii) The 1,2-diphenylethane is formed by two $C_6H_5CH_2\bullet$ radicals joining together in a chain termination step. ✓

$C_6H_5CH_2\bullet + C_6H_5CH_2\bullet \rightarrow C_6H_5CH_2CH_2C_6H_5$ ✓

(iii) The reaction mixture will contain some hydrogen chloride, unreacted methylbenzene and some polysubstituted methylbenzene. The following procedure should be used to obtain pure chloromethylbenzene:

- Wash the reaction mixture with water and then with sodium carbonate (or sodium hydrogencarbonate) solution ✓ in a separating funnel, collecting the lower organic layer ✓ each time.

☑ This removes hydrogen chloride, which is a reaction product impurity.

- Dry the organic layer with anhydrous calcium chloride. ✓

- Filter off the calcium chloride and distil ✓ the clear dried organic layer. Discard any unreacted methylbenzene that distils over at around 111°C and collect the fraction that boils over between 177°C and 181°C. ✓

☑ The polysubstituted methylbenzenes have a higher boiling temperature and are left behind in the distillation flask.

The periodic table

Group

Period	1	2													3	4	5	6	7	0
1	1.0 H hydrogen 1																			4.0 He helium 2
2	6.9 Li lithium 3	9.0 Be beryllium 4													10.8 B boron 5	12.0 C carbon 6	14.0 N nitrogen 7	16.0 O oxygen 8	19.0 F fluorine 9	20.2 Ne neon 10
3	23.0 Na sodium 11	24.3 Mg magnesium 12													27.0 Al aluminium 13	28.1 Si silicon 14	31.0 P phosphorus 15	32.1 S sulfur 16	35.5 Cl chlorine 17	39.9 Ar argon 18
4	39.1 K potassium 19	40.1 Ca calcium 20	45.0 Sc scandium 21	47.9 Ti titanium 22	50.9 V vanadium 23	52.0 Cr chromium 24	54.9 Mn manganese 25	55.8 Fe iron 26	58.9 Co cobalt 27	58.7 Ni nickel 28	63.5 Cu copper 29	65.4 Zn zinc 30			69.7 Ga gallium 31	72.6 Ge germanium 32	74.9 As arsenic 33	79.0 Se selenium 34	79.9 Br bromine 35	83.8 Kr krypton 36
5	85.5 Rb rubidium 37	87.6 Sr strontium 38	88.9 Y yttrium 39	91.2 Zr zirconium 40	92.9 Nb niobium 41	95.9 Mo molybdenum 42	[98] Tc technetium 43	101.1 Ru ruthenium 44	102.9 Rh rhodium 45	106.4 Pd palladium 46	107.9 Ag silver 47	112.4 Cd cadmium 48			114.8 In indium 49	118.7 Sn tin 50	121.8 Sb antimony 51	127.6 Te tellurium 52	126.9 I iodine 53	131.3 Xe xenon 54
6	132.9 Cs caesium 55	137.3 Ba barium 56	138.9 La lanthanum 57	178.5 Hf hafnium 72	180.9 Ta tantalum 73	183.8 W tungsten 74	186.2 Re rhenium 75	190.2 Os osmium 76	192.2 Ir iridium 77	195.1 Pt platinum 78	197.0 Au gold 79	200.6 Hg mercury 80			204.4 Tl thallium 81	207.2 Pb lead 82	209.0 Bi bismuth 83	[209] Po polonium 84	[210] At astatine 85	[222] Rn radon 86
7	[223] Fr francium 87	[226] Ra radium 88	[227] Ac actinium 89	[261] Rf rutherfordium 104	[262] Db dubnium 105	[266] Sg seaborgium 106	[264] Bh bohrium 107	[277] Hs hassium 108	[268] Mt meitnerium 109	[271] Ds darmstadtium 110	[272] Rg roentgenium 111									

Key:

Relative atomic mass
Atomic symbol
name
Atomic (proton) number

Elements with atomic numbers 112–116 have been reported but not fully authenticated

140.1 Ce cerium 58	140.9 Pr praseodymium 59	144.2 Nd neodymium 60	144.9 Pm promethium 61	150.4 Sm samarium 62	152.0 Eu europium 63	157.2 Gd gadolinium 64	158.9 Tb terbium 65	162.5 Dy dysprosium 66	164.9 Ho holmium 67	167.3 Er erbium 68	168.9 Tm thulium 69	173.0 Yb ytterbium 70	175.0 Lu lutetium 71
232 Th thorium 90	[231] Pa protactinium 91	238.1 U uranium 92	[237] Np neptunium 93	[242] Pu plutonium 94	[243] Am americium 95	[247] Cm curium 96	[245] Bk berkelium 97	[251] Cf californium 98	[254] Es einsteinium 99	[254] Fm fermium 100	[256] Md mendelevium 101	[254] No nobelium 102	[257] Lr lawrencium 103

questions & answers

PHILIP ALLAN
UPDATES

Edexcel Chemistry

(2nd editions)

George Facer

- Comprehensive coverage of the specifications for AS and A2 Chemistry

- Worked examples to show chemical processes

- Examiner's advice to consolidate understanding of key topics

- Short-answer questions and unit tests for quick revision and exam practice

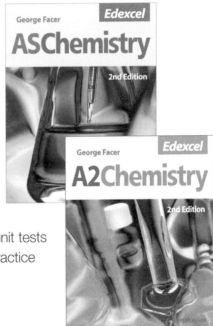

PHILIP ALLAN
UPDATES